ATLANTIS

ATLANTIS

FACT: Two Colonies Inhabit the Atlantic & Pacific Oceans

Written and Edited by
ROSEMARY KLEM

Zodbooks

First Edition Published as an eBook December 17, 2012 by

PO Box 117 Riverstone NSW 2765 Australia
Website: www.zodbooks.com
Email: info@zodbooks.com
Phone: +61 (0) 419 204114

This Edition Published May 2013

ATLANTIS
FACT: Two Colonies Inhabit the Atlantic & Pacific Oceans
Copyright © 2012 by Rosemary Klem

All rights reserved. No part of this book may be used or reproduced in whole or in part, copied, scanned, uploaded, electronically shared, resold, stored in a retrieval system, or transmitted in any form or by any means – electronic, mechanical, photocopying, recording, or otherwise – without the written prior permission of the publisher or the author.

Edited by Rosemary Klem
Cover design and artwork: Rosemary Klem © copyright
Interior design: Rosemary Klem
Photograph Jack Lord (2013) by Rosemary Klem / © copyright Dream Shots
Photograph Jack Lord (taken in Italy after being in a coma) with permission of Jack Lord

Cataloging-in-Publication details are available from the National Library of Australia

ISBN: 978-0-9871167-2-7

Contents

The First Word 9
Introduction 11

BOOK I

A MOTHER SHIP IN THE ATLANTIC OCEAN

1   Telepathic Communication 17
2   Cybernetic Organism 21
3   The Lift 31
4   The Bedroom 33
5   The Bathroom 43
6   The Wardrobe 49
7   The Study 53
8   The Dining and Living Rooms 55
9   A Description of the Mother Ship 59
10 The Living Center of the Mother Ship 65
11 Cleaning 69
12 The Life Cycle: Health / The Family Unit / Childhood and Education /
    Marriage / Death 73
13 The Governing Authority 85
14 The Kitchen 89
15 Food Tablets 93
16 The Balcony / Garage 99

17	The Main Shopping Center / A Small Boutique / The Physics of Clothing and Shoe Adjustment	103
18	Robots	113
19	The Command and Navigational Centers / The Physics of Space Travel I	119
20	Flying Saucers / The Physics of Space Travel II / Security and Combat	125
21	A Trip in an Indoor Flying Saucer / Traffic Lanes	133
22	The Beach / Sexual Attraction	137
23	Dolphin Surfing	143
24	The Bird Life	151
25	The Streets	153
26	A Book Printer	161
27	Abductions	163
28	Reporters and Cameras	165
29	Antarctica	169
30	A Visit to a Recreation Area at the Lake	173
31	A Visit to Two Small Retail Shops	179
32	A Tribute to the First Visitor from Terra	185
33	Guyd Kicks the Bucket / My Legacy	193
34	A Visit to the Museum	201
35	Lunch at a Restaurant	205
36	A Visit to the Art Gallery and Library	209
37	A Historical Look at the Empire of Atlantis	217
38	A Farewell Celebration	221
39	My Body Wakes from a Coma	229

BOOK II
A PERMANENT COLONY IN THE PACIFIC OCEAN

Introduction to Book II 233

1 My Astral Travel Visit to the Colony in the Pacific Ocean 237
2 A Big Brother Society..241
3 A Description of the Colony...247
4 Fishing Areas 251
5 The Bermuda Triangle / The Physics of Space Travel III 253
6 Construction of the Colony..261
7 Farms in the Pacific Ocean / Agricultural Science /
    Tablet Manufacturing / The Physics of the Food Tablet 265
8 Economy and Trade / Legal System 273
9 The People..279
10 The Streets, Homes, and Life in the Colony....................................283

Conclusion 289
Also by the Author..293

The First Word

What we don't know
Is more than we know
And stranger than we can imagine

Atlantis . . . the "lost continent" made famous through one historical source: the Greek philosopher, Plato (360 BC). The mere knowledge of its existence has been solely preserved in a few passages of his writing. What we cannot perceive is that Plato's account goes beyond the relevance of preserving the idea that there was once an existence of Atlantis, which would otherwise have been lost to us; it carries the clue to the question of who we are, where we come from, as well as why we are here. There is no perhaps about it: Atlantis is even the key to unlocking the mystery of our destiny.

In his works, Plato presents to us only a vague notion of Atlantis. His facts are inadequate at answering the basics of why, how, and who. He only gives us the clue to where. For him to answer the former – that is, why, how, and who – was an impossibility in his time. Words to describe Atlantis would not have existed even to this great thinker. And, just as knowledge has suffered throughout the ages at the hands of vested interests, to whom knowledge and truth are a poison, his knowledge would have been venomous, more so than those truths for which the great Socrates lost his life at the hands of his accusers.

Plato started something that he could not finish in that life span. This book finishes what he started. What a story there is to tell! One that answers the questions we have asked about our existence: questions on our past, questions on our future. With this book comes a new beginning for man on Earth.

Introduction

Few words are necessary to introduce this non-fiction book. Atlantis is not a myth. Nor is it a lost continent. Atlantis — which represents a superior race of humans that originated from another planet, and which established us on this planet — presently exists on two fronts: a mother ship settled in the Atlantic Ocean — which is the basis of Book I; and a growing colony settled in the Pacific Ocean — which is the basis of Book II.

In any work on Atlantis there is often a reference to the Bermuda Triangle, for good reason. What connection the Bermuda Triangle has to Atlantis is intriguing. What is unmistakable is that this book will on the one hand wind up the endless speculation surrounding this mystery, and on the other hand add a compelling twist to it, to the extent that researchers will be tempted to undertake a new wave of exploration of the region.

Additionally, this book provides us with clues to solving some age-old mysteries, such as the Egyptian pyramids and the great flood. Lost in the riddles that speckle our yesteryear are specks of our tomorrow. In our examination of the mother ship as well as the colony, we will be looking at those specks — that is, at the blueprint of our future.

The peoples of this globe share a commonality that goes beyond our borders, our customs, our hopes, and our beliefs, for, we are the "adopted children" of a superior race of humans, to whom our very existence as we know it is attributable. These humans have been here with us, closeted, throughout our entire history, with an end in mind, of which we will learn.

From the outset, it is understandable that one may be skeptical and question the veridical (truthful) nature of this work. All anyone can ask is that you take a journey in the words that fill it. What will be apparent is that no author is so imaginative that he can make up a work of this kind. What will also be apparent is that science fiction, theorists, and futurists have been inadequate in depicting the future.

There are two parts to Atlantis. Book I is a recount of the visit made by Jack Lord in 1959 when he was in a coma. On this occasion, he was able to stay in the mother ship, which is in the Atlantic Ocean, for a lengthy period. The details of his coma and how he came to visit the mother ship are not touched upon here; they are in the biography of his life.

Book II is a recount of the visit made by Jack Lord on December 8, 1994, at the age of fifty-five, using the technique of astral travel. On this occasion, he was able to make a brief visit to the colony in the Pacific Ocean.

Because astral travel has been germane to the attainment of the knowledge embodied in part of this work, the following is a brief overview of the human body and astral travel; what we can expect from this overview is to demonstrate to ourselves that astral travel is real and possible. We should note that the human body is not complete without a metaphysical component – most of us would understand this as being the soul. In certain circumstances, the metaphysical body can exist without its physical body. One of these circumstances is death; another is astral travel. Many of us have never heard of astral travel, while some of us have some vague or ill-conceived notion of it. To astral travel is to leave your physical body with your metaphysical body in a conscious and controlled way, which is similar to the near-death experience. Some of us will know that both the Russians and the Americans have been using a primitive version of this technique for spying purposes.

We should note that the technique of astral travel is being tossed around frivolously these days, and confused with vivid dreaming, which is not a consciously-controlled experience but a subconsciously-controlled experience. In other words, when you astral travel you are one hundred percent conscious, just as you are in wakefulness, and you control your movements and actions; when you vividly dream you are a maximum of eighty percent conscious, and your conscious mind has no control of your movements and actions. (The state of dreaming is explained in another book.)

Astral traveling is dangerous, and you should only consider it when you are aware not just of its risks, but of what you are actually doing. In some circumstances you can even time travel. Several rare humans throughout our history had this ability, one of whom was Leonardo da Vinci. By this technique, some humans were able to gain an insight into the future. (Only our ignorance can make us dismissive of what will one day be reality and common knowledge.)

The following is a technique that enables you to prove to yourself that astral travel is not a fairytale from a fanciful mind. Make sure you are alone in your bed when you attempt this, as you should not be moved while you are in an astral travel state. Make sure that you are physically healthy and possess an abundance of energy, such as aura. Make sure that you are emotionally well – that is, you are not stressed or in a negative state, such as feeling depressed. Make sure that you are not under the influence of liquor or any form of medication. Make sure that you are aware that there are dangers.

In bed, before you go to sleep, you should relax to the point that you are about to fall asleep. For most, this process is easier to attempt after several hours of sleep. By then you should be relaxed and possess energy that you do not normally possess when you first go to sleep. Do not have any thoughts on your mind. Imagine that energy in the form of aura is flowing throughout your body and concentrate on it. Picture it as red or blue. At the point that you are relaxed and feel as though you are about to fall asleep, imagine that your metaphysical arm – as opposed to your physical arm – is moving out of your body. Do not move physically. You should feel the manifestation of two physiological symptoms: a buzzing noise and a vibration moving throughout your body. This is the detachment phase. If you suddenly hear a voice in your mind telling you to stop, then heed the words and stop the attempt. You should also stop if you are too afraid of proceeding to the next level. The physiological symptoms you experience in the detachment phase should be enough to satisfy your curiosity that astral travel is real.

Should you continue to imagine that your metaphysical arm is moving out of your physical body, your astral body will detach. It is best to have your hand not by your side but under or near your forehead, and to be lying on your stomach, so that when you are out you will see your physical body below you. The initial shock and fear will cause you to return to your body instantly. If you are successful and plan to pursue this activity further, it is advisable that you know of all the dangers involved, and of how the metaphysical body works in a physical body. (These are also discussed in lengthy detail in another book.)

Finally, it is with a combination of my intuitive ability to construct a blueprint of the physics of some of the technology, and the narrative provided to me by Jack Lord, that I have written this book in the voice of Jack, to whom these events occurred. What you are about to read is a true story. The following details have been provided to the best of Jack's recollection, and to the best of our ability as laymen to interpret the physics of the advanced technology.

Let me say that interpreting the physics of the technology has been no easy feat, as it has involved trying to understand unknowns that are beyond the understanding of a layman – no doubt, physicists may have even felt a little challenged trying to interpret the physics! One can say this with an edge of cockiness, when you consider that physicists are so wrong on the most basic questions relating to the origin and nature of the universe. This tome provides a realistic blueprint for physicists to take us into the future.

BOOK I

A MOTHER SHIP IN THE ATLANTIC OCEAN

1
Telepathic Communication

In my metaphysical body, I appeared in a room. I was not alone. Present in the room was another human. Also present was the person who was responsible for my being there. The details of who this was, and how my being there was possible, are not a part of this recount, as they are the subject of another book. Nevertheless, this person introduced me to a man called Guydar (spelling may not be accurate), who became my guide during my stay there. I thought his name was odd, but, then, in view of the circumstances I was in, who was I to question what was odd? Guydar said, "You can call me Guyd."

I was introduced to Guyd as Jack Lord. This is noteworthy because, many months after this experience, I would find a new country to live in and become a citizen of that country. In my new start, I would officially change my name to Jack Lord, and this would be with no conscious recollection of my visit to Atlantis. When I was in Atlantis, I knew that Jack Lord would be the name I would choose; apparently, so did the person who introduced me to Guydar. In a metaphysical state, you know a great many things.

I understood that I was to stay in Atlantis while my physical body remained in a coma. It was up to Guyd to tell me when its condition changed and when it was ready to accept me back. When the time came, Guyd would take care of my return to it.

All of our conversations were by telepathic communication. All intellectually developed humans communicate by this means. Only a small percentage of human civilizations is inadequate, or incomplete, in the same way that our civilization on Earth is. There are two types of telepathy used by superior humans.

The first form of telepathy involves hearing words in your mind. The difference between a verbal conversation and a telepathic one is that you don't hear with your ears just as you don't speak with your mouth; the mind is the medium through which either task is performed.

What was interesting was that all of our conversations occurred in English. I was only fluent in one language, which was not English. Yet while I was in Atlantis I was well versed in English. I know that in one previous life I was fluent in English. After reading *The First Cause, Volume I*, we will understand that past lives are stored in the subconscious mind; in a metaphysical state (such as death or astral travel), there is no division of the conscious and subconscious minds as humans on Earth presently experience in a physical state. At first I assumed that this might account for how I was able to have such an intimacy with English. Eventually I learned that the model of body I was in has a component of its brain that interprets languages. It does not interpret in the same way that our interpreters on Earth do. It allows you to understand English, even if you have never spoken a word of the language before.

What is also interesting is that English appears to be not only a universal language spoken throughout the universe, but also the preferred language. Even though languages have their own process of evolution, that they are used elsewhere in the universe tells us that language is not something we have developed on our own without external intervention. This is the subject of another book.

The second form of telepathy is called "knowledge transference"; the computer equivalent of this is to download. You will experience an instance of knowledge transference at the moment of your death, for, this is when your life flashes before you. To your mind, time may feel "stretched"; yet this is all it is: a moment in which you feel everything that it took you a lifetime to experience and feel. In only one instant, a great deal of knowledge pertaining to your life, which has been stored in your subconscious mind, is downloaded into your conscious mind.

When a writer or poet feels an entire story in his mind in one "burst" of knowledge, then he has experienced an instance of knowledge transference. In only a flash, he will know all the details of that story. Again, this is because the knowledge has been downloaded into his conscious mind for that fleeting moment.

Often knowledge came to me by the method of knowledge transference. When knowledge comes to your mind by this means, you clearly understand everything about a subject. Now, to retrieve that knowledge and interpret it later are completely different challenges. And, then, to find words for someone else's "knowns" that are our unknowns is another matter altogether. It is all logical and understandable at the time the knowledge is downloaded into your mind, but for you to find the language to convert it to words is next to impossible when the words don't yet exist.

Knowledge transference involves using the "Universal Language," which is described in detail in another book. The Universal Language is a language of feeling, which can be interpreted into English, German, French, mathematics, art, music, the written word, and so forth. How to picture this is simple: when you have a word on the tip of your tongue, but you cannot think of the word, it means that you know the word in the Universal Language, but you have failed at that moment to convert it to your language, let us say, English. However, you do feel the word. If you are French you will feel the same word, but you will interpret it with a French word. The same if you are Japanese – you will find a Japanese word. If you are an alien, you too will feel the word in exactly the same way, but you will interpret it with a word from your alien language. If you are an intelligent German Shepherd, you will not be able to interpret it with a word, but you will feel the word and understand it. If you are an academic, you will probably use a sophisticated word, while a layman may not. The argument I am trying to make here is that the words and sentences I use are not always the direct words, sentences, or grammar used, but words according to how I have interpreted them within the limitations of my knowledge, language skills, and understanding.

2
Cybernetic Organism

At first, I thought that Guyd was human; he appeared human in every sense. I soon learned that this was not the case: it was inconceivable that beneath his human exterior was a cybernetic body. In terms of physical appearance, he appeared to be in his early thirties. His face was photogenic – movie material as far as I was concerned, and show business was my interest and career at the time.

Guyd said, "My skin is living tissue. We call this living tissue 'cybernetic organism.' This means that my body is half man, half robot. This living tissue, cybernetic organism, is composed of microorganisms, which are created. When an artificial intelligence is finished off with cybernetic organism, he possesses all the qualities that it takes to impersonate a human. For instance, a robot like me sweats and stinks; he bleeds; he even has bad breath. He even eats and drinks. Drinking water is necessary to prevent the skin from dehydrating. This means that with this body you have a need of discharging waste, the same as a human has. If you see a robot, you will never suspect that he is anything but a human." He tapped me on the shoulder and smiled, just as a parent does to a child when that child experiences or sees something new and often beyond his comprehension. It appeared to me that a highly sophisticated robot has been degraded to the status of a human just to be identified as one and to be indistinguishable from one.

He continued, "While you cannot at first glance distinguish between a robot and a human, there is a vast difference in the two. For instance, a robot is impossible to destroy. You cannot kill him. If you come close enough to him and know what to do you can disconnect him so that he cannot function, but there is no way you can destroy him, even with an atomic bomb. If you shoot him or even cut him he will certainly bleed, but all he has to do is pass his hand over the wound to aid in its repair, and it will repair almost instantly. Let us say that you put a hole through his head with a bullet. What will happen is that the damaged area will restore itself to its original form and integrity, to the extent that there will be no indication that there ever was a wound in the first place."

He demonstrated this to me by cutting open his arm. Blood was flowing from the wound for only a moment. He motioned his hand over the wound, which I learned was only significant in speeding up the healing process. The hand motion supplied additional aura, which is a healing agent, to the wound. I watched the wound close and the skin restore itself to its original state. Then Guyd simply brushed the dried blood away with his hand.

I learned that how this is possible involves some startling technology. Today, we know that atomic particles are not elementary, which means that they can be subdivided into smaller particles, which are protons, neutrons, and electrons. We also know that protons and neutrons are not elementary, which means that they can be subdivided into even smaller particles. Naturally, atoms are so small that we cannot see them with the naked eye, but they do exist everywhere. Everything you see around you, including you, is composed of atoms, and atoms are composed of subatomic particles. We have a long way to go before we can ever identify the family of subatomic particles. Right now, understanding the field of quantum mechanics (the study of subatomic particles) is to us what the unknown world was to Columbus before he set sail on his voyage of discovery.

By contrast, these superior humans have mastered the field of quantum mechanics to the extent that they can break down any matter state into its subatomic particle state, and then either reassemble subatomic particles into their original matter state, or create a new matter state from subatomic particles. They can use the atoms that are in the air to create a matter state. Interestingly, old stories tell us that this is what a controversial figure from our past could do. It is conceivable that a human can turn one fish into many fish: first, by copying the subatomic particle blueprint of the first fish; then, by using that blueprint to convert subatomic particles from the air into many fish!

One may think that while it is credible that technology can perform this, it is not credible that a human can. The answer to this may surprise us. Technology, at best, is only an imitation of what a superior metaphysical body is capable of, and, what a superior physical body with no restrictions (this last point is significant) is capable of.

If this technology seems impossible to us, it is only because we understand so little about the building blocks of matter. Yet once we have mastered this great unknown – the field of quantum mechanics – putting together a matter structure from subatomic particles will be as routine as assembling a model airplane. After all, everything you see is made up of an arrangement of subatomic particles. That arrangement is the blueprint. Once we know how to reverse the structure of matter – that is, break it down into its subatomic particle constituents – then it will not be long before we figure out how to arrange subatomic particles into structures of matter. This means that with this technology we will have the ability to create anything, even a physical clone of ourselves, simply from subatomic particles taken from the air.

How a robot is able to repair an injury is simple, and it involves subatomic particle technology. A robot has his subatomic particle blueprint registered in his brain. When he is injured, his subatomic particle blueprint has altered. His body is programmed to revert from any changed subatomic particle state (i.e. an injured state) to its original subatomic particle state (i.e. an uninjured state).

While this may seem a fantasy to some of us, to the rest of us parallels in our human body will immediately be apparent. We only have to look at some of the marvelous healing processes of our body. Our cells contain programs that instruct our body to perform certain actions. Growing a second set of teeth is one example. What stem cells within our body are capable of when "switched on" is another example. Our energy is low and our immune system is weak at our present stage of intelligence. Energy and the strength of the immune system are proportional to intellectual development. Just as the strength of our immune system and the level of energy we possess proportionally increase with our intellectual development, so a time will come that programs in our bodies will be switched on in such a way that our bodies will not "decay" or "corrupt" as easily as they are presently designed to.

Subatomic particle conversion technology plays a significant role in the lives of humans in their society, of which we will learn as we proceed through this book.

It must be said that solving the subatomic particle riddle will lead to drastic changes to our society. If we are not careful, this technology has the potential to see the decline of society. On Atlantis, technology has not clashed with society in such a way; it has only enhanced it.

What I found interesting was that Guyd could see my astral body. You could be mistaken for thinking that an astral body does not have a presence in a physical realm. However, a metaphysical body possesses energy in the form of aura, which can be visible to humans. Furthermore, being invisible does not restrict a non-physical entity from communicating with a physical entity. Telepathy has no borders. This means that if one is developed enough, one can detect, feel, and communicate with a metaphysical presence.

In a business-like way, Guyd said, "You are going to be given a body like mine for the duration of your stay here. This body and your metaphysical body are going to be synchronized as one."

We entered into a room, in which there were several people. "These are robot technicians, who operate the technology. First, you're going to go through this doorway into a corridor, which is divided into five different rooms. The door will automatically open and then close behind you. The walls will look like glass. You will see on the floor, in the center of the room, a black oval spot. You stand on it, and a fog will emerge from it. This fog will encompass you from the ground up. From a distance, you will appear to be in a cylinder of fog, but the cylinder of fog that will appear to be around you is not a physical structure, so you can lift your arm and penetrate through it, but that is not advisable.

"The fog you will see creates cybernetic organism; when it is on you as a finished product, you will look the same as any human that you see here amongst us or on your planet. With this body you will have feelings as well. You will also have blood circulating around you because your skin is a living organism, and it possesses blood just the same as normal skin does. The difference is that if you injure yourself, within seconds the wound will close, no matter how damaged it is, and you will feel neither the healing process, nor the pain. The wound will automatically seal itself and there will be no residual effect to indicate that there ever was a wound, as you saw in my earlier demonstration.

"There is no negative energy within you while you are in this body. (No negative-part, which is explained in another book. In short, every human is deliberately corrupted with negative energy.) In addition, any negative energy that tries to penetrate this body will just bounce from it, which means this body is protected from this energy, and other energies in the universe that can destroy a human. There are only a few that can destroy a human in this body, like gamma rays, but this body will long in advance warn you of those places that such danger lurks. All the humans you see that work in the domestic industry here are like me. Under our skin, we are technological. And we do not have souls. In your case, while you are here you will have the body of a robot, but you will have a soul.

"In the first room, a thick layer of skin will stick on you. When the fog disappears, you walk through the door to the next section. Once again, the door will close behind you automatically; then you go and stand on another black oval spot on the floor. The same thing will happen. This time a thinner layer of skin will be added, which is about three millimeters thick. You repeat the process in each section. Each section will fine-tune your body. You will pass through five different sections; in the fifth section, a blue fog will purify you so that no bacteria have invaded your blood.

"By this stage you will have a technological body, which has a replica of the human heart in it. This heart, made from a hard substance, acts as a pump, and it pumps blood through all the parts of the skin and brain. Your body will be a replica of your physical body, as you know it. Although it will be technological, your skin will be cybernetic organism, which will give you a realistic appearance."

Guyd then told me to go through the process. I felt neither scared nor apprehensive, as you should in such a circumstance. But, then, I was not in possession of a physical body to feel such emotions. In my metaphysical form, I felt ineffable (unspeakable) joy and possessed abundant energy, which, within the constraints of a human body, are what you just do not feel and possess respectively.

When I approached the door, it slid open sideways. During my stay, I never once had to touch a door through which I walked. Through that doorway was a corridor, and at its end was a see-through door. I could see similar rooms and doors beyond. I stepped on the black oval spot, which was two and a half feet in diameter, and stood still. Above it was a corresponding black oval spot. In no time a fog, which was the color of Caucasian skin, discharged from the black oval spot I was standing on. The fog engulfed me as it rose to the black oval spot on the ceiling. As soon as it touched it, I felt a cold sensation. I was careful not to move. As quick as the fog appeared did it disappear. I knew that with its disappearance it was time for me to move through to the next section.

I walked up to the see-through door, and it automatically opened and closed as I passed through it. I stepped on the black oval spot that was on the floor in the middle of the room. This time a different colored fog appeared, and when it reached the top black oval spot, I felt little by way of sensation.

I entered the next room and continued the process. In the third section a different colored fog appeared. So with the fourth section. As with the previous two occasions, I felt little in terms of sensation when the fog reached the black oval spot on the ceiling. However, I now had a full body. In the very first section I had an appearance, but not until the fourth section did I have a full body. Before I went through this process I could not see myself at all, as I had no physical body.

In the fifth section, a faint blue colored fog appeared. On this occasion, as the fog was rising around me, I felt as though I were having a shower, from the ground up. I knew that this process was disinfecting me, by eliminating any harmful bacteria that may have invaded my body. Although I had the sensation of needing to dry myself when the fog disappeared, I was actually dry. Yet I felt fresh, as if I was brand new.

At the end of the fifth room I encountered the final door. This time the door opened to a brightly lit room. Here, there was a mirrored wall, which I found confronting because it revealed my reflection. My metaphysical body had been "dressed" in a physical body. What this physical body did for me was give me a physical presence in the physical realm.

Until this moment, I had not seen anyone other than Guyd and several robot technicians. However, in this room there were other corridors leading into it, from which I saw several humans exiting. In view of their responses, I could tell that they were in a similar circumstance to me, and that they were just as scared at the sight of their reflections as I was of mine. I saw one fall onto his backside in shock. All were cursing and swearing aloud. Even though I could not understand their language, I knew they were cursing and swearing because I was able to read their minds.

Later, I would come to learn that these people had ended up there after being captured in the astral plane. In my case, I was brought there. They were astral travelers from other planets, who had ended up there either by mistake or intentionally. Astral travelers who are captured are usually kept there for experimental purposes for a certain time before they are sent back to a place familiar to them. If they were to have any recollection of the experience, it would be in the form of a nightmare or a dream.

The reflection I saw in the mirror was of me, skinny and nude. It was easy to tell that I had a physical form because my emotional state had altered once I was in a physical form, in that I felt scared and insecure.

The body I possessed was not the same as the body we possess. It was the body of a robot with the skin of a human. This means that my brain and all my organs were artificial. While I was there, my soul was able to fuse with artificial intelligence. It should be noted that humans from their world do not have this type of body. They have biological bodies that live and die, just as we have. While the robots in their civilization have feelings, they do not have a soul, and they do not live and die. To maintain their existence, a technician services them.

A voice in my mind told me to go into a room through the door that was on my left, where I would find something to wear. The room was empty, apart from a walk-in wardrobe. There were several suits hanging in it, not to mention half a dozen pairs of shoes on a shoe rack. There was also a chair, as well as a foot stool.

The first thing I did was choose a suit. There was not a choice of sizes in any of the clothes or shoes. I was surprised that they only had the one size. At first, I concluded that they knew my size, but that was not the case. When I put an item of clothing on, the strangest thing happened. Initially, the item was not a perfect fit. Yet, within moments it molded to my body so that it became a perfect fit. For instance, initially the suit jacket was in places too big and in other places too small. The sleeves were too long, as was the length of the jacket. However, the material automatically adjusted itself so that the jacket became a perfect fit. The length of the jacket and sleeves shortened. Areas that were too tight loosened. Areas that were too loose tightened. With every item of clothing I put on this occurred: underwear, shirt, trousers, and socks. Astonishingly, this also occurred with the shoes. I slipped my foot into a shoe. Instantly, the shoe molded itself to my foot so that it was a perfect fit. It was extraordinary!

I found it significant and interesting that one size of clothing can adjust to fit the shape, size, and height of a person's body, because it means that there is only one standard size of clothing sold in their society, and it meets everyone's physical needs. The exception applies to children's clothes, which come in smaller sizes; however, these also adjust in size to fit the child's body shape. In a later chapter, this book will analyze the physics of how clothing is able to tailor to the needs of the body in this way.

I dressed in a suit and tie, and then walked out of the room. Guyd greeted me and said, "Come with me, and don't be afraid. We can go wherever you want to go, but I'll first suggest where. This is where your tour begins."

I cannot explain how I felt at that moment. I was speechless. My overall experience up to then was stupefying. That I should discover firsthand that death is not the end and that there is a metaphysical existence beyond the physical existence was mind-blowing, but that I should witness firsthand our "future" and our "parent" civilization was miraculous, for want of a better word.

⁌

When I had crossed a fortified border to escape from the communist country I had grown up in, a barrage of machine-gun fire pierced me; as a result, I lost about half of my blood. The result of this trauma to my body was that I ended up in a coma for fourteen days (which was when I made my visit to Atlantis). That was when I experienced my first out-of-body encounter. While out of my body, I met "someone" who introduced me to the world of astral travel. Most of my subsequent astral travels during my lifetime were short encounters. On that first occasion, my out-of-body experience was a lengthy one. I did not have to worry about being "pulled" back to my physical body, which was what happened on almost every other occasion. When you astral travel, the amount of energy your physical body possesses dictates how long your metaphysical body can remain out of your physical body. My circumstances were unusual, and a little complex, so this condition did not apply to me when I was in a coma. As long as my physical body remained in a coma, my metaphysical body was able to stay out of it, with no constraints imposed upon my metaphysical body while it was out.

They say that from every bad there can be found a good, if you choose to seek it, or let it seek you. My coma was my good fortune. For it gave me an opportunity I would not have had. My stay in Atlantis was for around thirteen days. I rarely got tired while I was there and I did not follow a night and day routine of sleep and wakefulness. I only went to sleep when I felt that I had to sleep, for, there was too much for me to see and experience. From the outset, I was so lost that I did not know what to see or think first. One thing after another overwhelmed me, and there was so much that I could have asked and should have asked.

My time there seemed to pass so quickly. In this recount of my stay there, I cannot put what I did in a day-to-day itinerary, so the following is a rundown of what occurred to me, and it is not necessarily in the order that I present it. I made many visits to shopping centers, to businesses, to manufacturing industries, and so forth. I visited a lake, a waterfall, and a beach. I was driven in a flying saucer and in a shopping trolley. I was shown the store of the future, agricultural science of the future, and cybernetic technology. I was given the basic details of space travel and propulsion physics, of inter-dimensional travel, of computer technology, of food technology, of cleaning technology, of the governing structure of a future society, and of intelligence and education. Now that I look back, I wish I had the wisdom then that I have now. I would have asked additional questions, particularly on elementary physics. I was too overwhelmed with seeing, absorbing, and experiencing – in my own peerless way. Perhaps when I see it all again, I might describe some things differently or from a different perspective; seeing something new can be hard to digest, especially when you only see it once, and then attempt to describe it over half a century later when you are not in good health and feel like you have death lurking around the corner.

3
The Lift

Guyd presented himself in a business-like manner the entire time that I was with him. Initially, this led me to suspect that robots have no sense of humor. I understood that a man in his position has many things on his mind. Humoring me would have been the least of his concerns. I took a different view; I went out of my way to make him smile. I felt familiar with him, from the moment I met him – probably a little too familiar that I could not contain my natural propensity to want to say something stupid. For instance, I asked him to smile like me. I demonstrated a fake smile and paraded my teeth. My motive was to test his reaction: to investigate the psychological nature of robots. I knew all the human responses I could get, and I knew what buttons needed to be pushed to get those responses, but I was in unknown territory, and I wanted to test the responses of a robot. I certainly gave Guyd reason to show me a side to him that I wasn't sure he even had; what I found was that a robot will not respond in a negative way to any of your comments. He exhibited none of the traits a human from Earth with poor self-control exhibits.

In response to my fake smile, he imitated it and asked, "Are you satisfied with my smile?"

I could not resist the leading question and said with a smile, "You look like a jackass with your nice big teeth?" What made me think of a jackass were his large, perfectly aligned teeth.

"In this case, let's go to your quarters." With a serious face, he stepped in front of me and then smiled. I couldn't see him smile, but I knew he was smiling because I had read his mind. He didn't want me to know, but one thing you cannot do is hide your thoughts when telepathy is a method of communication.

We headed to what would be my living quarters for the term of my stay there. When we arrived at the lift, the doors instantly opened. Guyd said, "The lift reads your mind. Every brain here contains an identification code. In identifying you, the lift's computer knows where your quarters are. The lift will only allow you into your quarters if the computer has identified you as living in those quarters. It will not allow anyone else in."

Once the lift opened directly into my quarters, he added, "Every unit has its own private lift. This means that there are no public lifts. When the lift is not in use it disappears. Although there is nothing to identify that there is a lift, you know where lifts are. When you require one, you can make it appear by mere thought. This means that when it is not used it disappears from sight."

To this day, the lifts confuse me when I try to picture them. It is illogical and physically impossible that every unit has a lift. The only way that this makes sense is if the lift operates on a different dimensional plane. This is not as absurd as it sounds.

4

The Bedroom

I stepped into an empty room. All I saw were bare walls, empty space, and a marble floor. The walls were painted in a light gray color that was tinged with olive. The color scheme was pleasing to the eye and peaceful to the mind.

The first thing Guyd said was, "All the units are soundproof, and have an in-built warning system in case of disaster; in such an instance, everyone aboard is transferred to safety, such as on flying saucers."

I walked around a little, and could see several other rooms, which were also empty. What I expected to see were rooms that were fully furnished in a futuristic style of décor. I was baffled when he added, "This is where you are going to be staying while you are here. Everyone has the same living quarters as this one; the only difference is the number of bedrooms. Come with me." We stepped into a room. "This is a bedroom for two people."

I was told that the sleep cycle of the people is much the same as ours; that their brains are in principle the same as ours, in that the model is based on a conscious and a subconscious mind. The difference is that the door between their two minds is permanently open, which allows them to use a large proportion of their brain. They dream just as we dream, although their dreams are vivid, which is because their bodies possess an abundance of energy.

Guyd made a holographic screen appear in the air, and then he asked me to choose a style. The screen, which this book calls the "décor menu," is a standard feature of the unit. It displays images of different styles of décor. The bedroom décor menu displays different bedroom styles. I remember seeing written in English, on the bottom of the first image, "Style 1." The images change, illustrating a range of bedroom styles, including styles from different periods of our history. Some styles are simply weird. I had the option of choosing different sized beds, different room arrangements, and all that. The décor menu offers a style to cater to every taste. The interesting aspect is that you can change your décor to suit your mood as often as you want. One day you may want French décor, while the next day you may want one of those weird styles.

Guyd asked me to stand under a black oval spot that was on the ceiling, just off-center in the room. There was a corresponding black oval spot on the floor. We stood near or on that spot. He said, "Now, you choose the style that you want, and then give a command to the computer to furnish the bedroom in the order you want it furnished. Do you want the bed to come out first? The wardrobe stocked with ready-made suits in different sizes – in case you get fat?" He didn't laugh when he said this, but I did, aloud. Evidently, this was his idea of a joke, because, as we know, clothing does not come in different sizes.

At first I found it difficult to get used to not speaking or laughing vocally; it took me a little while to use telepathy. Besides, whatever I said sounded funnier by voice. On occasions, I deliberately spoke verbally to add humor to the situation, and humor usually came about by my saying or doing something stupid; it has always been my natural inclination to say or do something stupid.

"Or, do you first want the viewing screen to appear, in which case you can watch some scenery, or have some relaxing music playing? This is a good way for you to go to sleep. You can ask for anything, in any order you wish."

I thought for a moment and then said, "I've never been in a place like this. What do you suggest I ask for first?"

"Wise decision. This is how you address the computer. You say, 'Computer, set up the bedroom the way you feel best suits my personality.'"

I repeated those words in my mind.

Within seconds, there was a slight noise. Then a blue fog filled the room. During this process, Guyd told me that the blue fog does most of the work in the flying saucer, and that bacteria cannot survive it – not even bacteria that cultivate in water. This process took minutes. When the blue fog disappeared, I was so in awe of everything I saw, and tried to absorb it all to the extent that I didn't even try to figure out how things worked. With the disappearance of the blue fog came a refreshing scent. What first struck me was the sound of a waterfall. On a large screen that occupied a great portion of the wall before me was a waterfall.

As for the rest of the bedroom, it was fully furnished in Queen Anne style, which, of all the styles in the décor menu, I loved most. Without even having to tell it, the computer knew that this style best suited my personality. Of everything in the room, the waterfall fascinated me the most. It reminded me on Niagara Falls, but it was so much nicer. In the universe, there are scenic attractions that rival those we have on Earth. The scene on the screen appeared real, in every sense. I felt I was there, in person. Emitted from the flora growing around the waterfall was a strong aroma, and from the scene itself, a crisp, salubrious air.

The technology involved in the appearance of the furnishings is interesting; it is based on subatomic particle technology. Each style in the décor menu has a corresponding subatomic particle blueprint. When you choose a style, conversion technology reads the subatomic particle blueprint of that style. According to the blueprint, subatomic particles are arranged into a matter form – in this instance, the furnishings. Just as the furnishings appear on command, can they disappear on command. When the furnishings disappear, there is a reverse process of conversion, where a matter form – the furnishings – is broken down into subatomic particles. Any matter state can be broken down into its subatomic particle state. Information in terms of the blueprint of this subatomic particle state can be translated into signals that are capable of traveling at extreme speeds anywhere in the universe.

In response to seeing the décor of the bedroom, I said, "This room looks much the same as the rooms from where I come. Victorian grandeur."

"It is popular here, especially for honeymooners."

Before me was a double bed covered with a luxurious bedspread. The bedspread was floral in design, with colors of olive, gold, and red. There were pillows on the bed, two bedside tables, a bed head, and a chest at the end of the bed containing bed accessories.

In a unit such as this, you will never have to make your bed again if you don't want to, nor will you have to wash the bed sheets. Every time your bedroom reappears from its subatomic particle state, it is brand new. Sometimes, nevertheless, some women do make the beds. They do this because they want to experience a traditional female role. Some people keep the same style of décor and don't convert the bedroom back into a subatomic particle state. However, they do not wash their sheets. Later, we will discover their method of cleaning.

There were rugs around the bed. The color of the marble floor matched the walls, only the floor was a darker shade. The ceiling matched the shade of the walls.

There was a one-inch gap between the cornice and the ceiling. Guyd said, "Fresh air filters through this gap. There are return-air outlets on two sides of the room, but you cannot see them. You cannot hear any sound created by the flow of air, nor can you feel any flow of air. You only feel that the temperature is right for your body, and you never sweat or feel cold. The computer reads your body temperature once you enter the room; accordingly, it adjusts the temperature of the room. Your body, then, is the thermostat that determines the temperature of the room."

Ornaments decorated the bedside tables, and there was a vase of fresh flowers on one. The vase looked as though it were made of crystal, but upon closer inspection, I realized that it was made of diamond. There were bedside lamps that looked no different from our lamps. What I found startling about the light bulbs was that their power was generated not by electricity but by subatomic particles. The following is a layman's interpretation of the technology, which, no doubt, will give physicists an opportunity to have a bit of a giggle. A light bulb is able to capture subatomic particles from the air. Once captured, there is an attraction between positive and negative energies, and the ensuing interaction creates a glow that is capable of burning for ever. Imagine if static electricity were captured to generate power. The two examples offer a basis of comparison.

The lamps turn on and off by both touch and telepathy. However, if you fall asleep with the lamp on, it will automatically turn off when you fall asleep. The intelligence in your room is able to determine when it needs turning off. This applies to the television and other devices in your unit.

With the exception of the wall with a screen depicting a waterfall, and the semicircular wall chandeliers on the walls, the walls were without adornment. The wall chandeliers were positioned all around the bedroom walls, and they matched the central chandelier. Across the top of each chandelier was a strip of pure gold.

Diamonds rather than crystals hung from chandeliers. Two small candle lamps lit each wall chandelier.

Guyd said, "You can decide which chandeliers you want on and which you want off. All you have to do is point a finger and then without moving your hand, make a downward motion with that finger. The computer understands the gesture. However, it is not necessary to use your finger; you can use your mind. When you are teaching someone how it all works, as I presently am with you, it is easier to demonstrate it by gestures. When you don't want the lights there at all, then you make a stop sign with your hand and push it forward. The computer will understand the hand gesture and the wall chandeliers will vanish.

"While the furnishings are arranged according to the standard style chart, not only can you determine whether the furnishings are present in the room, but you can determine the placement of those furnishings. For instance, you can rearrange the placement of the vase with flowers. Perhaps you just want a bed and nothing else. Perhaps you prefer not to have wall chandeliers. You make a decision and tell the computer of that decision."

Guyd proved my earlier thoughts on robots not having a sense of humor wrong, for he made a second joke: "You don't even need to have any furniture in the room if that is your wish. You can just sleep on the floor. That will be fine. But I suggest you be careful exactly where on the floor you decide to sleep, because if you accidentally dream about the code, the computer may pick it up and your furniture might just appear, in which case you will be accidentally knocked about. While it never happens, it is capable of happening. This is why we stood on or near the black oval spot that is on the floor and ceiling of every room."

He was right in that there is a possibility that the furniture may knock you about when it appears. During my stay, I could not help but be tempted to experiment with making furniture appear and disappear. I was young, and I had grown up in conditions of poverty under the rule of communism, so I found the technology a great deal of fun. I was like a little kid who suddenly had everything. There was a sense of excitement and childishness in my behavior. I don't think they have ever experienced having anyone quite like me there before. In those moments of indulging myself in stupid experiments, I had an immaturity that not even the children in their society have. Nevertheless, I concluded from these experiments that if you tell the computer to do something stupid, the computer responds to you, telepathically, "This order does not compute." On one occasion, I tried to stack all of the tables and chairs from the dining room on top of one another. The computer did not find it amusing. It asked me, "What is the purpose of your request?"

"I just want to see if the computer is smart enough to do something stupid!" I answered.

"The computer does not compute such orders," it responded. The computer was always polite to me, and logical. I even found the occasion or two that it had a sense of humor. Below are two instances, which will give you an idea of the caliber of intelligence of their computers.

There was an instance when Guyd and I were window-shopping. Nearby were some young girls who were around sixteen years of age. Out of nowhere, the computer telepathically said to me, "Your zipper is open." Guyd burst into hysterical laughter, while the girls closed their mouths in horror and ran away, as though I were a monster.

Once he managed to contain his laughter, Guyd asked, "What did you do to the computer to make him embarrass you in this way?"

"Nothing," was my reply, which was not exactly the truth.

"Well, you must have done something to offend him."

There were many occasions that the computer embarrassed me publically. For instance, when I purchased some chocolate from a chocolate shop.

"Watch out, computers love chocolate!" Guyd said.

The sales girl just opened wide her beautiful eyes when she heard the computer's remark: "Just be careful about your wife; she might punch you in the nose if she finds out that the chocolate you just bought is a present for another woman."

Guyd could not contain his laughter. Feeling embarrassed, I gave him a dirty look. It was obvious that he and the computer were conspiring against me. I didn't need Guyd to work against me, when the computer was doing a good enough job of embarrassing me whenever it could. It regularly made remarks to me and those I was in the company of, or those who were nearby and able to pick up the telepathic transmission. You cannot defend yourself against this type of attack or invasion of privacy.

These examples tell us much about the computer network. A computer is much the same as a robot, just packaged differently – it has feelings, intelligence, and a sense of humor. If you forget its physical nature, you would think it were human.

Additionally, it knew my every move and every thought. In the final analysis, the computer appears to be an extension of your consciousness, of which you have no conscious awareness until a computer communicates with you. This means that the computer is tuned in on all consciousnesses – that is, on all the humans and on all the living creatures – and even on all the robots. What amazing technology!

You may have formed the impression that consciousness exists in the walls, in the computers, and even in the air. In a sense, there is some validity to this, for, the whole mother ship can be considered conscious. It is as much a living mother ship as our planet is a living planet. (There are comparisons in the two, in ways that we would not understand at this instant. Had you read *The First Cause, Volume I,* the comparisons would be obvious.) This is truly remarkable, and hard to picture. This book has thus called the computer network a "conscious computer."

Let us compare the conscious computer to a human brain. The human brain has consciousness, which is also known as the soul. However, it also has a second consciousness, and sometimes even more (depending on your intellectual development), which is the voice that we often hear – the one that sometimes guides us. This is called a "first-part." (More on this is explained in the above-mentioned book.) In exactly the same way, the freethinking, conscious computer has its own attachments. While a first-part is a personal attachment unique to each human brain, nanorobots and consciousnesses from the governors are attachments to the conscious computer. They are everywhere on the mother ship, and are in the millions. In a remote way, they are tuned in on and linked to every single life form, including robots, on the mother ship.

What should be obvious is that these people have mastered robotics in a way that is beyond our wildest imagination. They have incorporated robotic technology with nanotechnology. Nanorobots don't take the form of a human. It is not necessary for them to appear human in their cycle of existence. If you take the mind of a robot, and all the abilities of a robot, and if you package these features in a miniaturized state, then you have the answer to much of their technology, including how the conscious computer is an extension of the consciousness of a human. Nanorobots have a brain, telepathic capabilities, and all those capabilities a robot has. Additionally, a nanorobot uses antigravity technology for its travel capabilities. This all means that a nanorobot is freethinking, and is able to assess every circumstance independently and make a decision. A nanorobot has designated functions, and its entire existence is geared around performing those functions. All the while, the conscious computer not merely knows exactly what all the nanorobots are doing, but sees through their eyes.

Later on in my stay I was shown other units, just to see some bedrooms of children. Each child has his own bedroom, and in each child's bedroom there is a study. In the study, I saw wall units with books, and millimeter thin monitors. The monitor is all there is to a computer. A computer has no wiring, no keypad, and no mouse, which makes sense when you consider that it operates by touch screen and telepathy.

The furnishings in a child's bedroom reflect the personality and age of the child. I saw three-D pictures of monsters on their walls – when you see one of these you instantly feel that the monster is about to grab you.

As we know, the three-D pictures, the television screens, and the computers are all the same thing. What is interesting is that what is on them interacts and engages with the viewer. The best way to describe this is by the following example. If a child is watching a show about, or has a three-D image on the wall of, a monster, a holographic replica of the monster's head and hands can on occasions stretch out of the screen up to him, in a move that takes him by surprise. It does not entirely come out of the screen, but parts of its body stretch out. This suggests that the program is aware of a human's presence and is able to interact with him in this way. Obviously, the intelligence behind this is the conscious computer.

You can even engage with people on television programs. You can have a debate with someone on a show. If you make that someone mad, he can partially come out of the screen and lunge at you as though he were about to punch you, and he will go within centimeters of you. No doubt, you will have fled by then! This shows us that television shows, computer programs, and three-D images are combined with three-D holographic technology. The intelligence involved makes the programs or images you are watching unpredictable and interactive.

Of the toys the children have, what caught my eye most were the dolls that are capable of having a real conversation with you, and the computerized animals with the same capability. One child had a dog; however, this dog did not drool and it did not shed hair. Another child had a lizard. One had a monkey. From what I saw, pets look no different from and behave identically to their biological counterparts; you simply cannot tell the difference between a real pet and a computerized one.

5
The Bathroom

Guyd said that every bedroom has a bathroom en suite. There is another bathroom, almost hidden from view, in the corridor, situated just before the bedroom. I did not even notice it until Guyd pointed it out. This is a medium-sized room with a toilet and a vanity unit. Its style of décor is determined by your choice of living room décor. The bedroom en suite automatically appears when the bedroom décor appears, and its style of décor corresponds with the bedroom décor. I became curious and walked into the bathroom for the first time. Bathrooms are large, about double the size you find in a standard home today. In the entrance of my bathroom, which had a three-way design, was a vanity unit with a beveled mirror, which had a square three-inch diamond frame that reflected a magnificent array of colors. I never believed that something so majestic could exist. Guyd said, "The material looks like real diamond, and, as you can see, it is used extensively in our décor. You will see it in chandeliers and on shop fronts. The material is unbreakable and practically lasts for ever."

While I could see a toothbrush in a golden toothbrush holder that had space for several toothbrushes, I could not see any toothpaste. It was only because I was a visitor that I had this in my bathroom. No one needs toothpaste; no one even needs a toothbrush. Why no one needs a toothbrush or toothpaste is that teeth are automatically cleaned by the blue fog when a person has a shower or washes his face, as we will soon learn. Furthermore, their teeth are perfectly aligned; this means that they do not have spaces between their teeth. Their teeth grow perfectly by a natural process, which means that no one needs dental care, let alone an orthodontist, in their society.

Their bathroom functions in the same way that our bathrooms function. Yet there is a vast difference. For a start, you will not find toilet paper in their bathroom. You will find a toilet, a vanity unit with a mirror, shelves and cupboards stocked with items for personal hygiene, a shower, and a spa. Unbelievably, you will not find the one thing that you would expect to find in a bathroom: water.

As a part of the vanity unit, there is a small hand basin. There is nothing extraordinary about this at first glance, until you try to wash your hands. I noticed that they don't have taps to turn water on and off. There is also nothing extraordinary about this in this day and age. Sensors detect when your hands are under the "water" outlet.

The outlet does not emit water. In place of water, a blue fog is emitted. Guyd said that the blue fog representing water is the same as the blue fog that I encountered when I went through the process of receiving my body. It is the same blue fog that reconstructs the furniture. Even though this is described as a blue fog, we must note that it is only about four percent visible to the naked eye. The blue fog that replaces water is pale in color, and is like a light fog.

When you wash your hands, you simply place your hands under the outlet. After sensors detect the presence of your hands, the outlet discharges a blue fog. It is not necessary to rub your hands or use soap. There is no soap in a bathroom of this kind. The blue fog does all of the work. While the blue fog is not water, it certainly feels like water. There is no difference between how the blue fog feels in this instance and how water feels. Indeed, the blue fog feels exactly like water when it is meant to represent water.

If you want to wash your face, then you splash the blue fog on your face, in the same way that you do with water. It feels no different. Only, you don't have to dry yourself afterward. When the blue fog disappears, you are automatically dry. This is because you were never wet in the first place. The blue fog gives you the sensory illusion that you are wet.

One would be right in suspecting that the blue fog is part of an intelligence that is able to construct matter from, or deconstruct matter to, subatomic particles.

In the bathroom, the blue fog possesses the knowhow to deconstruct matter. Additionally, the blue fog is able to determine what is considered as relevant matter and non-relevant matter. Ostensibly, the technology behind how the blue fog is capable of this may seem vague. Freethinking intelligence is the factor at play. The freethinking intelligence involved needs to assess each circumstance and make a determination. The blue fog on its own is not capable of this. Nanorobots must be the answer to the intelligence involved. Wherever there is a blue fog, there are nanorobots.

The nanorobot in the blue fog in this instance has the job of determining what is foreign on the skin. Once it has made an assessment, the nanorobot breaks down anything foreign on you, such as dirt, into its subatomic particle constituents. The blue fog goes into a drain-like facility, which is a part of a recycling system that travels throughout the mother ship. You know that your hands are clean when the outlet ceases to emit a blue fog.

The door of the small room with a toilet was open, but once I stepped in, the door automatically closed behind me. The toilet looked like a modern version of our toilet. You have to sit on a toilet seat and pass waste in the same way that we do. The difference is that there is no water in the flushing system. There is no flush button. The blue fog replaces water in the toilet, and it converts your waste into subatomic particles. As we know, in every blue fog there is a nanorobot. When you sit on the toilet seat, the blue fog will sense your presence; as soon as you excrete any waste from your body, the blue fog will travel to the source of the discharge, and the discharge will be converted into subatomic particles. This means that you will never see your discharge. In addition, as stated earlier, there is certainly no such thing as toilet paper. There is also no such thing as smell.

You can watch television while you are in the toilet. Furthermore, in every public toilet there is a television screen, and you are able to choose whatever channel suits you. In public toilets, you do not have to worry about picking up anything hazardous to your health! And you certainly can sit on the seat of a public toilet. Toilet seats are about two inches thick and soft to sit on. They are never cold.

I took a closer look at the shower, and noticed hot and cold water buttons. The principle of the blue fog replacing water applies to having a shower and washing your hair. If you were dirty before your shower, you will no longer be dirty after your shower. If your hair was oily before your shower, it will no longer be oily after your shower. Once again, you don't experience water in a shower; being in a blue fog shower feels no different from being in a water shower. Even though you can still feel wet, you are completely dry. Your hair is even dry. Therefore, it is not necessary to dry yourself with a towel after you have a shower.

Another part of the shower process is that when the shower is complete and switches off, you may hear a subtle bell sound in your mind. The telepathic bell sound is commonly used in different areas of their life, such as in manufacturing. There are different tones that are symbolic of different things.

It should be of no surprise to learn that the spa also has a blue fog in place of water. In it, you feel as though you are in warm water. If you feel the need to make the so-called water warmer, you ask the computer to do this. With your eyes closed, you will not be able to tell the difference between the kind of spa we presently take with water and the kind of spa they take with a blue fog. With your eyes open, you will see a spa full not of water but of blue fog. To the mind, the blue fog feels like water.

It should be noted that while you are in a spa or even taking a shower, the nanorobot in the blue fog doesn't just remove sweat and dirt from your body, and oil and dirt from your hair. It also looks for such things as bacteria and bad cells, which in our world lead to diseases, cancer, and so on. This means that the nanorobot goes through all the molecules of your body; it even goes through your teeth, and, in the unlikely event that it finds a flaw, it will rectify it or eliminate it. In the "pure" environment that the mother ship provides, the possibility of a human having a flaw in his body is small; therefore, it is doubtful that a human will ever have a disease or sickness.

In the instance you step out of the spa or shower and feel wet, then you have a towel hanging nearby, with which you can dry yourself. When you take a towel, a new towel appears in place of the original towel. Most people don't bother with a towel because after a while you become accustomed to not feeling the sensation of being wet. The sensation goes away because you are actually dry. Feeling wet is only a condition in your mind. Thus, the intelligence of the blue fog plays on your mind to create the illusion of wetness. Towels are more likely to be used on the occasion that you did not bring a gown with you into the bathroom. Your modesty may then cause you to cover yourself with a towel.

Near the towel is a dirty-laundry basket. This is a small square outlet on the wall, which is not obvious until you want to toss your clothes, shoes, or even towel into it. This is when the outlet will appear. Intelligence in the dirty-laundry basket will know of your intentions; to correspond with those intentions, the outlet will appear for you. A blue fog will also appear in the basket, and whatever you tossed in will instantly disappear. By this stage, the outlet will have closed and you will only see a wall. In the case of my bathroom, there was a three-D image of a red rose with a stem and green leaves.

The science behind the dirty-laundry basket is subatomic particle conversion technology. For the contents of the basket are broken down into their subatomic particle constituents. First, intelligence in the basket senses when you are going to place an item into it. Once an item is in the basket, that intelligence knows what that item is. Anything thrown into the basket, such as a towel, clothing, or shoes, is converted into subatomic particles. At the back of the basket is a small section that is the size of a business card, and it has small holes in it. These holes are connected to the recycling system.

An additional process occurs in the dirty-laundry basket: when wardrobe items are converted into subatomic particles, information relating to the subatomic particle signature of each wardrobe item is determined. This information is sent to the subatomic particle converter in the wardrobe. Based on this information, the wardrobe subatomic particle converter converts subatomic particles into the same wardrobe items that were broken down when they were put into the dirty-laundry basket – with one exception: the items are clean and look as though they have been ironed. In the case of your shoes, if applicable, they are polished. Everything then finds its way to its rightful place in the wardrobe. More on this is explained in the next chapter.

6
The Wardrobe

I was curious about the wardrobe, so I stepped back into the bedroom. The appearance of the wardrobe was not immediately obvious to me. All there was to indicate the presence of a wardrobe was a bare wall in the room, with a blue-gray light emanating from the gap between the cornice and the ceiling. I was told that the wardrobe only presents itself when you request it.

The standard design I chose creates one wardrobe in a double bedroom, which is shared. You can modify this and have two: one for each partner. Their wardrobe looks nothing like our wardrobe, and it is not just a place to hang clothes. It is also a laundry. In their society, there are no such things as washing machines and clothes dryers. There is no such thing as ironing clothes. There are no clotheslines, no pegs, and no washing detergents. What will be a delight for many to know is that there is no such thing as doing the laundry at all!

The wardrobe has two parts to it: a front section as well as a back section. Sensors trigger the doors to open sideways only when you come close to the wardrobe and intend to use it or request it to appear. The doors then disappear somewhere in the side walls. When the doors to the front section open, all you see are three bars that stretch across the width of the wardrobe, and these bars are the size of a finger. They are all in a line, at the same height, in a similar way that the strings on a clothesline are. Only, in this case there are three rows. They are a light blue-gray color. They are up high, so that they present no obstacle to you when you walk under them to go to the back section.

Holes on the bars form the basis of a magnetic system. Perpendicular to the bars, beams of light are emitted from the holes. These beams of light extend to the holes of the nearby bar. These lights are also a blue-gray color, and they effectively do what a peg does. There may be shirts hanging together in the space between bars, and each shirt will hang front on, parallel to the bars, connected by two blue-gray light beams. In other words, there is a beam of light penetrating through each side of the shirt, through each shoulder, for instance. In the same way that we have shirts hanging side by side in a wardrobe on coat hangers, they have a shirt hanging in front of another shirt, with a small space between shirts so that they do not touch. The beams of light going between bars go through all of the shirts. Each beam is the magnetic basis of what holds the shirts in place.

The front section of the wardrobe is usually empty. When you walk several steps into the wardrobe, you encounter another set of doors. These doors are closed until you step up to them. Your clothes are hanging behind these doors. In this section you will find additional bars, which are identical to the ones in the first section of the wardrobe, and which function in the same way. Thus, there are no coat hangers in a wardrobe of this kind. The clothes hang in groupings of similar items in an orderly arrangement. Shoes are stacked on shoe racks. There are shelves for items of clothing that are not hung, such as undergarments.

The first section is generally not your storage section. This has a different role to play. Let us say that you want to change your clothes. You approach your wardrobe and the doors will open. Then what you have to do is toss your clothes into the dirty-laundry basket that is in this section of the wardrobe. As soon as you throw your clothing and shoes into it, a blue fog will appear in it, and then the items in it will disappear. What is interesting is that the items will then reappear in the back section of the wardrobe; the result is that they will be placed in their allocated sections. You should have noted that items are not sent away for dry cleaning or polishing – whatever may be appropriate. That said, items of clothing do not hang in their dirty state – that is, in the condition they were in when you tossed them into the dirty-laundry basket. The items of clothing will be clean and look as though they were dry-cleaned. Shoes, if applicable, will be polished and placed on a shoe rack. However, an important consideration is that not any of the items will be brand new after they go through this process. They will remain in the condition – whether worn out or brand new – that they were in when they were tossed into the dirty-laundry basket, apart from being clean, or, in the case of shoes, polished. Therefore, if an item is worn out and almost ready to be thrown away, that is the condition in which it will reappear in the wardrobe. While the beam that pegs the clothes to the bars also keeps the clothes fresh and clean once they are there, it is not responsible for the initial cleaning process; for the cleaning process occurs in the dirty-laundry baskets.

This sounds impossible. It sounds fantastic. How can dirty items go into a dirty-laundry basket and then wind up clean in the wardrobe? If we ponder long enough on the question, we will discover the simplicity of the technology. We know that an item of clothing is broken down into subatomic particles. So too is any foreign matter on the clothing, such as dirt or stains. Whereas information on the subatomic particle signature of the clothes is registered, information on the subatomic particle signature of the foreign matter on the clothes is not. This way, when the information is sent to the converter at the back of the wardrobe, the item of clothing is then converted back into its matter state, only clean, and looking as though it were ironed.

What this tells us is that a nanorobot in the blue fog in the dirty-laundry basket identifies what your item of clothing is and what foreign matter on that item of clothing is.

A nanorobot performs another function in the dirty-laundry basket. It needs to make a decision on whether an item of clothing is wearable or not. When clothing becomes old or worn, which includes being ripped or damaged, the nanorobot will recognize the condition of the clothes before they are converted into subatomic particles. It will determine whether an item of clothing needs to be disposed of. What will then happen is that the item of clothing to be disposed of will be converted into subatomic particles and cease to exist. Not only that, the nanorobot will know whether you want that item of clothing replaced. It will have read your mind. Knowing that you want a replacement, it will order one on your behalf; a new replacement will be delivered to your address and it will end up in your wardrobe, and you will not even know it. Naturally, the replacement is charged to your account.

How the product goes from the shop to hanging in your wardrobe is an interesting process. Information on your purchase is transmitted directly from the shop to your wardrobe as signals. It is much the same as sending an email. The signal contains the subatomic particle blueprint of the item. Once the wardrobe receives the signal, it uses the attached blueprint of the item to construct it to its matter state. This suggests that many of the purchases you make do not need to be taken from the store with you when you purchase them.

As stated earlier, subatomic particle conversion technology exists in the back of the wardrobe. It is only when the wardrobe is closed that the converter operates.

Let us say that the wardrobe subatomic particle converter has received the subatomic particle blueprint of a jacket that you have just thrown in the dirty-laundry basket. All the converter has to do is create the jacket from subatomic particles and then make it appear in the jacket section of the wardrobe. The puzzling aspect of the technology involves just how the item is able to look as though it has just come from the dry-cleaner. Understanding this process itself is not as daunting as it seems. How your clothes go from looking worn in the dirty-laundry basket to looking ironed and crisp in the wardrobe involves a nanorobot. When the nanorobot breaks down clothing into subatomic particles, it is able to make subatomic particle adjustments to correct the state of the clothing. The nanorobot then sends this information to the wardrobe converter.

As a final note, if you want to hang an item of clothing in the wardrobe yourself, then you take it to the relevant section. Let us once again use the example of a jacket. You go to the jacket section. All you have to do is raise the jacket to the light beams that are emitted from the holes in the bar. These lights will magnetically attract the jacket. The jacket will then straighten and hang in place.

7

The Study

We went from the bedroom into the corridor and then into the study. I took the initiative and chose my own style from the décor menu. The décor menu offers many different styles from which to choose. I chose Queen Anne style. My choices were always either Queen Anne or French – I never went for the weird designs, but I saw several shops that did.

There was a triangular-shaped desk in a corner with a bookshelf above it. On the right of the desk was a wall unit, crammed with books. Positioned at right angles were two sofas. There was a coffee table, an end table with a diamond vase containing fresh flowers, along with a second reclining desk chair. On one corner of the wall was an oil painting of the master himself: Leonardo da Vinci. Guyd told me that it was painted by one of their artists. In another corner was an oil painting of the Mona Lisa. He also told me that family portraits, in the form of oil paintings, often hang in the study.

At the time I was there, computers were not known to me, so the basic concept of the computer was unbelievable, let alone its telepathic feature. The applications of a computer with telepathic capabilities and intelligence comparable to consciousness are considerable. For instance, to write a letter, you don't have to write it word for word. There certainly are no keypads. As soon as you think about the general content of what you want to say, the computer will convert those contents into a letter. It will take on the role of a ghostwriter and editor, or perhaps even a secretary. A writer will never have to face the dilemma of finding a grammatical mistake, or the dreaded missing word, in his published work, as he does now, despite the fact that he may have read his manuscript thousands of times, in an attempt to avoid this very thing from happening!

The computer monitors are thin. They have three-D technology. They also operate as television screens. One program that they have can zoom in on any location on Earth to the extent that they can view the activities of an ant in your backyard. They use this technology to observe not just our planet, but other planets. With this technology they can even view underground civilizations, such as the one on a planet in our own neighborhood. It may interest us to know that they can view us at any time of the day or night. They have a live feed on us – and there is no censorship, apart from certain lewd acts. This means that they can be watching what is going on outdoors anywhere on the planet, in anyone's backyard, at any time of the day. The public has an ability to monitor your yard twenty-four hours a day. There is no privacy outside of the home. Inside the home is another matter, unless you have been "tagged." Some of us fit into this category and provide a "live feed" of our lives to others out there. More on this is discussed elsewhere in this book.

8
The Dining and Living Rooms

Once in the living room, Guyd said, "Don't forget to stand under the black oval spot." I did not need reminding. He added, "Go ahead, do what you want to do."

Once I chose my style from the décor menu, a blue fog appeared. It only took several minutes to disappear. With its disappearance came fully furnished lounge and dining rooms.

I was told that once upon a time all the styles were designed by interior designers, and my thoughts were on how good they were at their job. Just as I did for the bedroom and study, I chose Queen Anne style furnishings for the lounge and dining rooms. I looked around and what first caught my eye was the dining room. There was a large dining table for eight. One thing that most impressed me was how glossy the marble floor was. It reflected the dining suite.

Guyd said, "The floor is not slippery at all. While it appears to be marble, it is actually a lightweight material – ninety-nine percent lighter than true marble. Everything in a flying saucer follows this principle of being made of lightweight material that imitates the original. In a vessel that travels through space, it is essential to minimize weight."

I noticed that table settings for eight were set on the dining table, as though a meal were about to be served. Guyd said, in response to my thoughts, "The dining table can extend to seat additional people. All you have to do is request a larger table with the appropriate number of chairs. Conversely, you can cut the number of chairs to accommodate an intimate seating arrangement. What you see is included in the standard design. It is your job to adjust the seating number to suit your guest list. This way you will only have settings for the number of people you require. In addition, you can request no settings on the table at all, or you can request table settings that suit the meal you are sitting for. It is up to you to suggest what is to be taken away, or what is to be added."

I picked up a knife and looked at its handle. It was gold, while the blade was stainless steel. I picked up a glass, which appeared to be made of diamond. Guyd said, "All the glasses for drinking are unbreakable. You can drop them and they will never break."

Hanging from the ceiling above the dining table was a diamond chandelier. Guyd said, "It never loses its sparkle – the blue fog keeps it looking brand new."

There was a bar, fully stocked with beverages. It was only because I was a guest that it was fully stocked. The standard design does not include food or beverages. I certainly did not expect superior humans to drink alcohol. The difference, I learned, is that while they do drink alcohol, their alcohol is unlike ours, in that it does not have any intoxicating side effects – that is, you cannot get drunk from drinking it. This is not entirely true, because you are able to feel its intoxicating effect if you let yourself, but this is only an illusion you create in your mind. This means that there is no alcoholism in their society. The possibility that people can be tempted to behavior that is destructive does not exist. Furthermore, the intoxicating effect is mild, and it does not allow the negative aspect of a human to overwhelm him.

The lounge room had an exquisite burgundy, quilted leather Queen Anne Chesterfield sofa, with two matching single sofas. They all had a high, quilted back with side wings. Beside the sofa were end tables. In front was a coffee table. Centered on the coffee table was a vase that appeared to be made of diamond, with fresh flowers – not just any flowers: the fragrance was unlike anything we have on Earth. The scent was hypnotic.

There was a large diamond vase in one corner of the lounge room; it too had fresh flowers in it. A large chandelier hung from the ceiling. Not any of the lights that burned emitted heat of any kind; they were interesting in the sense that they had captured the aura and essence of a crisp air that comes with a spring day, if this makes any sense. It is hard to describe.

There were six three-D television screens spread across two walls. There were three on one wall and three on the other, all aligned next to the other with about six inches of space between screens. No pictures decorated the walls. The television screens were about five feet wide by three feet high. I was curious about having so many screens, but I was told that when your guests are over, or all of your family is together, everyone watching is able to see all six screens, which is a good thing. At first this may seem chaotic, until you remember that there is no audible sound. This way, each person can telepathically tune in on a different television screen and hence a different show. The sounds of the other shows will not bother you because you tune in on the television that you are watching. If you have kids, they can watch whatever they want. When you look at a screen, you telepathically tune in to that screen; all the telepathic transmissions from the other screens are blocked. You can even go from screen to screen and watch multiple shows at once.

As far as the size of the television screens is concerned, you can change the size and the number of them. You can have only one television screen fill an entire wall. As mentioned earlier, the television screens have three-D technology. Three-D glasses are not necessary. Guyd told me that many people with phobias are too scared to watch certain programs. Imagine what it would be like for someone who has a phobia of snakes to be watching a show on snakes. Then imagine what it would be like knowing that his computer knows about his phobia – we know what kind of sense of humor the conscious computer has!

Let us put ourselves in this viewer's shoes. Imagine that there is a scene with a snake in the show you are watching; the computer may just take the initiative and modify the program to make a holographic three-D image of the snake parading its fangs dart out from the screen right up to you. While it does not leave the television screen, it does extend out of the screen. In this instance, you are likely to jump up from your chair and run for your life. If you hear laughter in your mind, you can bet it will be your computer laughing at you!

In another sense, imagine a scene like Niagara Falls. Even though you are standing on your marble floor, or even sitting on your comfortable lounge, with three-D technology you are a part of the whole scene; you feel that you are there and are going to fall over the ledge if you are not careful. You are on the edge of life or death: on the edge of danger. Your nervous system jerks, and you panic; you even lose your balance. Then you realize how silly you are, because none of it is real. You will probably hear laughter in your mind, and this will be the laughter of your computer. Sometimes it intentionally creates an earthquake in the show or image to make you respond in this way.

9

A Description of the Mother Ship

Guyd asked me to join him on the sofa to watch television – not one screen, but all six. Guyd said, "If having so many screens is bothering you, then all you have to do is tell the computer to turn them off." I did not mind having six screens, because I was fascinated with the different perspectives of information each screen offered.

On one screen, there was a remarkable scene set in the Atlantic Ocean. What I saw was a spaceship sitting on what appeared to be a leveled-off mountain. Guyd said, "This is our mother ship, called, 'Atlantis Mark-VII,' which is where we presently are." The size of the mother ship is impossible to comprehend. While I am not one hundred percent certain of its size, I know that it is at least the size of Sydney and its entire metropolitan areas.

"The spaceship is situated below the ocean's equilibrium of pressure. The mother ship has no problem being beyond this level, but it is not recommended for your humans using your current technology."

On one screen I saw the mother ship as she must have looked when she first settled in the ocean. On another screen I saw the current scene, as she presently looks. Many fish were swimming around it. In its present location, it has been disguised to blend in with the mountain and seascape, so that if you ever try to look for it you will never detect it. Vegetation covers it. Skeletons of long-gone marine creatures that once roamed the oceans also cover it. There is a vast tableland nearby, brimming with marine life.

Guyd continued, "The most important defense a civilization of our kind has when it is living amidst a civilization of primitive humans is its ability to exist undetected. This is why it is impossible for you to detect us with any of your technology, such as radar. The mother ship automatically absorbs radar signals. Along with other defensive strategies, the mother ship is practically invisible and undetectable.

"Furthermore, the mother ship — as well as the colony that you have yet to see — has an additional protective screen to protect it from intruders of any kind. This protective screen involves a force field. There are five stages to the force field, the first being an antimagnetic field. If you are persistent enough to go past this stage then you are likely to be destroyed in the second stage. The first stage is hard enough to penetrate, let alone the other four stages. This means that the force field will disintegrate anything we screen as being an intruder, and it also means that no one undesirable can get through it. Living creatures that belong to the natural environment are not affected in any way; only what is deemed a threat is singled out. The screening process is not a manual one, but an automatic one, and is able to distinguish between what is a threat and what is not a threat."

Once again, Guyd tapped me on the shoulder, smiled, and said, "Don't worry." You have to be in my place to understand the shock I was in.

"The mother ship has a population of over one and a half million people. This includes men, women, and children. There used to be around two million residents, but many have emigrated to the surface of your planet and have assimilated with you. They can never be recognized as aliens." Guyd told me a great deal in relation to this topic, which will be the subject of another book.

The mother ship is divided into two sections: a top section and a base. This book calls the top section the "living center," and the base, the "control center."

It is in the living center that the occupants of the mother ship live. At the very apex of the living center is a "technological center." This section spins. The physics behind the action of spinning is the creation of gravity. It is only when gravity is required that this section is activated, such as when the mother ship is in space. This means that gravity is created on demand. Gravity is provided to the occupants of the mother ship so that they can enjoy normal living conditions — whether in space or in the middle of the ocean. This book calls this section of the technological center, "gravity production center I."

We will learn that there are two different sections in the mother ship that can independently create gravity: gravity production center I in the technological center, and "gravity production center II" that is located in the control center (the base) of the mother ship. It is essential for the mother ship to have two independent sources that are able to create gravity, just in case one were to fail.

In the technological center, there is technology to not only collect energy from the universe (from which nuclear fusion is created), but also emit the navigational laser locking beams. These technologies have relevance to the mother ship's travel capabilities; each is described elsewhere.

The control center (the base of the mother ship) is divided into three sections. The general population is not permitted entry into the control center – only a few designated personnel are. Most of the machinery that it takes to run the mother ship is located in the control center, such as the engines and motors to run the ship, the defense systems, and the computer network.

The third section, the bottommost part, of the control center is used to store energy. This book calls this section the "energy storage center."

The second section of the control center differs from the other two sections. This is gravity production center II, and it rotates. As mentioned previously, the spinning action creates gravity.

The first section of the control center is the heart and brain of the mother ship. In it are the "command center," the "navigational center," and the "weapons center."

At the base of the control center is a "spike" that extends from its bottommost point. The spike is about one hundred meters long in its present position. The purpose of the spike is to lodge into a structure, such as a planet. This not just anchors the mother ship in its location, but provides stability and balance once it is firmly lodged in that location. This means that when the mother ship lands on a planet, the spike penetrates into the surface for about one hundred meters, but the size is variable and depends on the environment. The spike is like a diamond drill, and it "drills" itself into the ground.

While the spike may be the main stabilizing force of the mother ship in a grounded position, on its own it cannot prevent the mother ship from turning. This is why, once it is securely lodged, many others are automatically released from the base in key locations. When the release trigger is activated, they drill themselves into the ground to the required depth. They can adjust in size according to the conditions and requirements. These are not anywhere near the same size as the central spike; they are around one quarter of its thickness. Nevertheless, they provide the additional stability necessary to anchor a vessel of this magnitude. This means that no natural event, such as an earthquake, can destabilize it once anchored.

When the mother ship leaves its location, all the spikes are recalled to their original storage location, including the main spike. None stays extended from the mother ship while it is traveling through space.

One thing about the mother ship is that there are no windows through which to look. The television screens depict a live feed of the scene outside the ship, even while it is in motion through space.

Around the exterior of the control center are flying saucers that are magnetically attached to it. Guyd said that there are over two thousand of them. There are two varieties: those used by the population of the mother ship, and the smaller fighter crafts, which hold up to three occupants. More on this is explained elsewhere in this book.

Whatever the technology involved, not just the mother ship but any spaceship of theirs is capable of becoming invisible, to the extent that no technology is capable of detecting it. This technology, among other technologies, gives civilizations such as this the upper hand over the negative civilizations that exist in the universe.

All of this knowledge was just coming into my brain as I was watching the television screens. Often I would see subtexts on the screens; the writing appeared to be changing in language. I even identified the English language. There was statistical information on the mother ship itself, and this included its size, its present population, and how long it has been lodged in its present location.

Guyd said, "We have been located here in the Atlantic Ocean for nearly twelve thousand years, after the time of Noah's ark and the great flood." He told me the story of Noah's ark. He told me about the Egyptian pyramids. Our ancient history made complete sense in that all the pieces fit into a puzzle that was complete. (This history is the basis of *The First Cause, Volume II*.) He said, "The pyramids were built to be indestructible. Our people built the Egyptian pyramids in the sizes they did to tell humans that there is someone else out there, that they are not alone. We even left traces of our history in the pyramids, including where we come from. You will find that your civilization has hardly explored the pyramids." Guyd told me some interesting facts relating to the pyramids that we will probably never figure out, but these, too, are documented in the book earlier mentioned.

"One day, soon, we are going to start a massive operation to change the genetic structure of your human civilization on the planet for the better, so that one day you become exactly the same as our humans are intellectually; this means that you will have a greater capacity to absorb knowledge, learn, and develop your brain cells.

"We are ninety-nine percent the same as you are on planet Terra, and our people are planning to colonize the surface of planet Terra in the future. While you may be like us, we are far in advance of you intellectually, which is why the time has come for another round of genetic change to humans on a mass scale. We cannot colonize the surface openly in this way until humans evolve to a certain grade of intelligence. When the time comes, our people will openly come to the surface to live. Our people look forward to this day, which will be the day that your people as humans have grown up." More on this is explained in detail in the book earlier mentioned.

An interesting point he made was, "Mary from the bible, a virgin woman, was a part of an experiment we undertook. We made her pregnant by the injection of an embryo, which we implanted in her womb. This was the final experiment of our genetic engineering program involving your species. We have been engineering your species in such a way that you become exactly the same as us." Now I understood that the Atlanteans, along with "another party," had an interest in and a connection to Jesus. That Jesus was a part of a genetic engineering experiment involving humans and the DNA of a superior race of humans is without doubt. (The role of Jesus, his life, and what he really did and said, are explained in another book.)

10
The Living Center of the Mother Ship

The living centre of the mother ship is massive. Because of the sheer size of the structure, you don't feel as though you are in a spaceship; you feel as though you are on the surface of a planet. To replicate the sun, there are several huge lamps fixed to the ceiling. Every evening they are dimmed so that the street lights take effect, which is the closest to night the population experiences on the mother ship.

I was curious about the ventilation. Guyd said, "There is a constant temperature maintained on the mother ship, and the air is purified. Warm air rises to the roof. Carbon dioxide is filtered out, twenty-five percent of which is then sent back into the ventilation system for the benefit of the natural environment, as the natural environment produces oxygen. This is why we have forests of trees. Some trees are as large as the trees you have on Terra.

"There is even a beach on the mother ship, along with a sea. The beach has surf, and it attracts quite a few people; swimming is a popular recreation." When he mentioned a beach, my first instinct was to think of sharks. In response to these thoughts, he said with a smile, "There are no sharks, but there are different species of marine life."

Everything you would find in a modern city you can find located in the mother ship. What I learned from this visit to Atlantis is that the accuracy of the future, as depicted by humans, is poor. The mother ship has been designed to emulate a real-life environment. Imagine enclosing an area the size of Sydney and its metropolitan surrounds, perhaps even a larger area, in a flying saucer. The environment on the mother ship is one that is capable of supporting life on a self-sufficient basis.

"The living center is divided into six floors. In the living center are the living quarters, shopping centers, and business and manufacturing precincts. On the bottom (first) floor, in the 'center stage' region, is the recreational area, with parklands, a lake, and the beach. There are many trees, rocky outcrops, and varying landscapes, just as there are on planet Terra. Long ago, the trees were planted for that purpose." Frankly, I could have confused what I was seeing with a location somewhere on Earth.

What was fascinating was that I saw birds flying around. Guyd said, "To make the environment as natural as possible, we have included animals, some wild, some tame. The wild animals we have are of the variety that does not fight. We even have insects." I found it funny that he added, "One thing we certainly do not have is woodpeckers – they are not welcome here! We only have creatures that are beneficial in some way. Because this is no Noah's ark, we have no need of two specimens of every species. Besides, we can recreate anything we want from subatomic particles. What we also don't have is wild creatures such as snakes, spiders, rats, mice, cockroaches, flies, mosquitoes, and other such nuisance creatures that you have to put up with on the surface."

Guyd tapped me on the shoulder, and all I could do was feel overwhelmed. I must admit, I was speechless much of the time. I was often too overwhelmed to ask questions. Often Guyd would just tap me on the shoulder and say, "Don't worry."

In understanding the design of the mother ship, we need to visualize it from an aerial perspective. In the center, on the bottom floor, we need to imagine a circle. This area has parks, a lake, and even a beach. There are also businesses and shops on this level. Above this circular region is nothing but empty space, and this empty space extends right up to the top of the sixth floor, so that the open space creates the illusion of a sky.

From the bottom floor to the roof, all around the outskirts of this imaginary circle – that is, the center of the mother ship – is a structure. This contains the living quarters, shopping centers, businesses, and manufacturing precincts. This structure extends out to the sides of the mother ship. Let us compare this to a cake. Now, if you take a cake and carve out the middle of it so that you make a big hole, and if you carve up the rest of the cake into portions, from the center out, as you would before serving it, then this will give you your portions. Now, we need to imagine that the carvings we made are roads. (This is a rough comparison.) In picturing these roads, we need to be mindful that these are not the same as our roads. They are merely wide areas of open space going from the bottom floor to the roof of the sixth floor. They have no need of physical roads – they have long ago done away with vehicles that require a surface to run on.

The living quarters of the inhabitants of the mother ship only exist on levels one to five. Every floor has its own levels of units on it. Therefore, if you live in a unit on level two, you may have several units above you and below you, not to mention beside you. The units are much the same as matchboxes on top of and beside one another. On some levels, mixed in among the units you will find shops and businesses. Some areas are designated as industrial and business precincts. The sixth level is strictly a shopping precinct. This means that there are no living quarters on level six.

There are divisions between floors. Two floors are divided by an area of space. It is in this space that there is a computerized network. The network controls everything computerized on that floor. The network is vast, and it is linked to the conscious computer. There are specialized robots, called "technicians," that fix and maintain the computer network. The computer network creates no noise whatsoever.

To access your quarters you use your very own indoor flying saucer, which this book has abbreviated as "IFS." This is not the flying saucer described earlier that is magnetically attached to the exterior of the mother ship. This flying saucer is used within the bounds of the mother ship. You even have a garage in which to park it. Because there is no public transport as we know it in their society, indoor flying saucers are the key mode of transport available to the population.

Unlike on Earth where you have cars priced for the rich and cars priced for the poor, there is only one model flying saucer. What is interesting is that in their society there are no rich and no poor, for, everyone is equal. Everyone has an IFS. What is more, no one has to buy one; everyone is automatically given one when he has his own living quarters. In other words, there is one IFS per household. More on equality, as well as on the IFS, is explained in later chapters.

11
Cleaning

The blue fog technology that was described earlier cleans everything on the mother ship – whether in the unit, on the streets, or in the shops. The blue fog removes all bacteria, viruses, poisons, and so forth. We know that nanorobots are the technology behind this. This means that rubbish never accumulates on the mother ship. If there is a spillage of some kind somewhere in a public area, the conscious computer will be aware of it, and a nanorobot will clean it up using particle conversion technology. It is not hard to imagine that the mother ship is always clean.

No one ever has to do the cleaning. There are no vacuum cleaners, no mops, no detergents or disinfectants, no polishers, and certainly no brooms. The cleaning process that goes on in your home is a manual one, which means that you decide when your unit is dirty and in need of cleaning. If you fail to activate the cleaning process, your computer will remind you to. Your home computer can detect when your home requires cleaning.

To clean your unit, you activate the blue fog, and it will clean everything, even a minute particle of dirt. When the cleaning cycle is activated, two outlets appear: one on the ceiling of the lounge room, which is the suction system, and a corresponding one on the floor, from which the blue fog is emitted. The outlets are the size of a ten-cent Australian coin. A blue fog will be emitted, and it will spread into all of the rooms of your unit. (This is also how factories and shops are cleaned.) The fog will spread everywhere in the unit, in the same way that the fog of dry ice spreads when you pour boiling water over it. Once the blue fog has spread to every room, it will start to rise. By the time it reaches the ceiling it will have almost disappeared. The one outlet in the unit will then draw all the fog that is in your unit into it. This process cleans your entire unit, and it only takes seconds.

The outdoor environment is treated in a similar way, and can be cleaned in a timeframe of minutes. Although its activation is at the discretion of the conscious computer, it is active around the clock. You don't ever see this process as it occurs. The blue fog in this instance is not visible as the blue fog is in a domestic or commercial environment, but it is the same. Why the blue fog is visible in a domestic or commercial environment is that it satisfies humans that a process has occurred. If the blue fog were not visible, you may not know that a process has occurred. This is the only reason it is pale blue in color.

The science behind the blue fog is that it transforms bacteria, rubbish, and all that into subatomic particles. The nanorobot in the blue fog is able to determine what needs to be converted into subatomic particles. You can be in the blue fog and it will not harm you in any way. On the contrary, all that can happen is that it will clean you and your clothes. Some people choose to leave their unit during this process, but some stay in it and feel the effects of having a shower.

The implications of this technology are so far-reaching that we cannot possibly grasp the full extent of them. One is that this civilization has solved the problem of garbage. These people don't dump their refuse, and they don't even have to find extra space on the spaceship to store it. Only a civilization in its primitive stage of evolution looks for places to dump rubbish, and thereby pollutes and disrespects not just its land, not just its oceans, but space itself.

What we may or may not even realize is how bad for us our techniques of dumping waste are. Our waste is the manufacturer of new bacteria, some of which are life threatening. Not surprisingly, this then means resources are wasted because we are forced to find new cures for diseases that should not have been created in the first place. Our resources could have been better spent. We have let ourselves fall into a cycle that never ends.

Frankly, we hardly crawl forward, and there are those who have a great deal to explain for our backwardness. In the eyes of civilized humans, we are going backward instead of forward.

On a spaceship, especially one with manufacturing facilities on board, you cannot have waste, and you cannot toss waste into space. When it comes to the manufacturing sector, manufacturers have special garbage bins located in the workplace. Some of these are built into walls. A sensor will detect when you want to place garbage in a bin, which activates the opening of a slot in a wall, for instance. Once you insert your rubbish, the slot closes and a blue fog appears. The blue fog converts any rubbish in the mother ship into subatomic particles. The blue fog then goes into an outlet that is a part of the mother ship's recycling system.

Garbage bins in the mother ship all operate on this principle. In the streets, you will see small bins. The lids have sensors so that they open when you are about to place rubbish in them. Once the lid closes, a blue fog appears, which instantly converts the litter into subatomic particles. These garbage bins are not large. Many just look like columns. The style of bin corresponds to the style of décor in which the street has been fashioned.

12

The Life Cycle: Health / The Family Unit / Childhood and Education / Marriage / Death

In any society, just as the advent of technology leads to the rise of one product or discipline, so it leads to the decline or demise of another product or discipline. This is evident with the blue fog technology. Blue fog technology has made some remarkable changes in a society, leading not just to the decline but to the elimination of so many products, industries, and disciplines. An entire society is reshaped with this technology. The technology is so far-reaching that we need to pause a moment and think just how far-reaching!

In our thoughts on the practical applications of the blue fog technology, we are unlikely to have imagined the decline of the following discipline: the area of medicine. For the blue fog technology has eliminated the need of doctors and medicine. Pharmaceutical industries have been wiped out, as has most of the medical industry. No one in their society needs a doctor. There are specialists in the field, technicians, who can take your physical body apart and put it back together again, just as a car mechanic can with a car.

Blue fog technology has been instrumental in the demise of diseases, viruses, infections, and the like, which can exist in you and kill you. In their cloistered society, the Atlanteans cannot develop sicknesses of any kind. The blue fog can remove from the body anything that does not belong in it.

When you combine this with their degree of intelligence, they have the ability to live to 1500 years of age. What their intelligence has to do with ageing is not obvious; it is explained in the coming paragraphs. When you have all the elements that are associated with an intelligent human, your body can enjoy perfect health, until some of your parts wear at around 1500 years of age.

The first thing that we may wonder about is how their bone structure can cope with such a long term of life, when our bone structure barely makes it to fifty in good condition. Their bone structure is different from ours. Genetically, they have altered their bone structure so that as it develops in childhood it toughens in a way that we cannot imagine. Their toughened bones almost replicate artificial bones.

Moreover, in their society there are no medical practitioners or specialists, let alone surgeons, as we know them. There are no hospitals. Yet there are occasions that someone may develop a health problem. For instance, you may cut yourself. In normal circumstances, you will see your wound automatically heal. Your energy is partly responsible for this, and your energy, as we know, is tied to your immune system. If we consider the case of Jesus and his ability to heal, we will see similarities. The only difference between the healing ability of Jesus and that of a person from Atlantis is that Jesus had a greater amount of energy, and a correspondingly higher grade of intelligence.

To explain this in simple terms, aura and a strengthened immune system are proportional to intelligence. The higher the grade of intelligence, the greater the amount of aura you possess. In the case of Jesus, he possessed the highest grade of intellect on the scale: Intellect Mark-7. The people of Atlantis are nowhere near his grade of intelligence, even though they have a higher grade than we have. They are not the most advanced humans in the universe. Advanced civilizations would have nothing to do with us, and nothing in common with us. The Empire of Atlantis shares a commonality with us, and this commonality is our golden opportunity: to one day become "one" with its people. Their commonality with us is similar to the commonality a parent has with his children.

Now, if your child cuts his arm, you are going to motion your hand over the wound. What you are doing is adding your energy to the wound. The bones, skin, veins, and so on, should automatically heal. In odd circumstances the child's energy might be compromised, which may result in the wound not healing, even in view of your energy being transferred to the wound. Your only option now is to go and seek the services of a technician. This is the robot that looks after, repairs, and services robots, along with all the technology on the mother ship.

In the mother ship, a permanent blue fog permeates through the air, and in its own way it contributes to the longevity of the human. This is slightly different from the general blue fog that is used in the processes we are so far familiar with, such as in the cleaning processes. Why this is different is that it is not visible to the naked eye. It is not visible only because there is no need to satisfy a human desire to see a process taking place. In essence, it is tantamount to an air purifier.

The family unit, as we know it, is the basis of their society. The husband is the breadwinner, and his weekly earnings are enough to provide a comfortable existence for the entire family. Poverty is not a word you can find in their vocabulary. Families never struggle. They never have to worry about how they are going to put food on the table for their kids, or how they are going to put a roof over their heads. For a start, there is no such thing as real estate in their society. When it comes to living quarters, they are the same for all. There are no mansions for some, and cardboard boxes or tents for others.

Essentially, this snapshot of a lifestyle demonstrates that everyone is equal, which immediately brings to mind the philosophy of communism. What makes this lifestyle work is that no human holds a position of authority or power over another human. The only time humans are not equal is when they do something stupid, in which case they may be physically eliminated. Their souls will then follow a normal judgment process, which is what occurs when we die.

This means that there is no prison on the mother ship. If someone has evil intent, he goes through a correction process. How his mind is corrected is by a process of upgrading – that is, his intelligence is upgraded in such a way that he will not have the same evil intents.

In view of our understanding of the word communism, this book cannot use this word without dirty connotations. In human hands, the hands of those in positions of power, communism has become a corrupted ideal. Regrettably, in a human society such as ours on Earth, corruption is synonymous with power, or control over others, which often equates to money. This is why a shocking proportion of those in positions of power on this planet has a fate none of us would wish for. What is this fate? To be deemed not worthy of being reborn again. That leaves two options available to a human when being judged for the merits of his life. (This is a jaw-dropping exposé, and it is the basis of another book.) That we are ruled in a corrupt way is why we have a pathetic standard of living, when compared with the standard of living we should have.

This book has thus replaced the dirty name of communism with "equalism." In a society of their kind, there is no corruption in the governing ranks. Ideally, humans do not value material things the way we do. They do not care to be better than others, have more than others, be richer than others, or have power over others. Fundamentally, this means that there is no greed, envy, or jealousy in their society, as we know it. They value the essential things of life, which are often lost to many of us. Overall, what is of value to them is their knowledge. In this sense, the competition of knowledge is respected.

In view of this, institutions or shareholders do not capitalize from others, particularly in times of distress. One day, in our world, there will not be a disparity in those who have and those who have not.

On Atlantis, many people work for a living, but they do not have to work if they do not want to. Having said this, an intellect of this caliber needs to be challenged, and work is central to challenge. This is why work does not need to be forced upon anyone. Even if you don't work, you are still paid a living. If the situation arises that there is a shortage of workers, the governing authority will call upon you to work if you are not already in the workforce. Moreover, you will not question this but take on the challenge with pleasure and satisfaction.

Every society must have an economy. One thing about the economy of Atlantis is that there is no such thing as unemployment. The economy revolves around business and manufacturing. We may find it surprising to consider that there are factories on the mother ship; this explains the self-sufficiency of the society. Many people are employed in factories; however, they are not employed as laborers; they have technical roles. In their society, humans never do menial tasks, whether that is in the factory or in the home. Robots do the menial tasks.

These are traditional people with family values, who take delight in following a traditional style of life. The mother does not work when she has a family. She stays at home and looks after the kids. She also looks after the husband. She has plenty of activities to occupy her time.

People enjoy the old-fashioned roles – those that long ago were written into our DNA. A woman, not a man, cooks the dinner, and she serves her husband at the dining table. She will dish out the food onto everyone's plate. Often she will set the table herself, even though she does not have to, as the table settings are a part of the standard design of the dining room and will appear upon request.

Some things are "heritage listed" in their society, if this book can use this term in this context. This is by choice. Heritage listing in this context is applicable not to bricks and mortar, but to roles and values. Women in their society love to cook the dinner and serve their husbands. They are proud of this role, and do not feel degraded or devalued in any way. Just as many women in our society today do not want to be liberated, so women in their society do not want to be liberated. Women are happy with their role, and take pleasure in their role.

If a woman does not want a traditional woman's role – that is, have a family, raise kids, or serve her husband, she will not get married, and she will not have kids. She will pursue a career or follow the path that motivates her, and she has equal opportunities in her career pursuit. Her intelligence, as well as capability, is equal to that of a man in most ways. Her performance is not in question. However, she is still a woman, and as a woman, she is the inferior of the species. Whether women like it or not, this is written in their DNA. Men and women are not born equal, and this applies to all women in the phase-3 cycle of human existence – the cycle in which we presently exist. (The phase-3 cycle is explained in detail in another book.)

We must remember that these people live for up to 1500 years, and in this period of existence a couple will never be able to have more than two children. They only have a small window in their life span in which to raise them and exist as a family unit. Therefore, in this limited time, they will enjoy the moment, and they will relish their roles as husband or wife, and parents, in the true sense of these roles.

In their society, men generally do not have a role in the kitchen. Men go off to do four hours of work each weekday. They work for fun. Similarly, the women do the household chores for fun. Men work to fulfill a challenge and avoid boredom. Working is just a way for people to challenge themselves. Most are inventors, engineers, innovators who better the society, or researchers, such as in the field of space travel. Researchers are always trying to exceed their speed of travel in the universe. There are always new technologies to challenge them.

※

An integral part of a child's life is his schooling. Parents don't wait until their child turns five years of age to begin his education. The greater part of the child's schooling occurs in the home. It is a one student, one teacher environment, and the computer is the child's teacher. For three days a week a virtual teacher teaches a child in the home. As mentioned in an earlier chapter, television screens and computer screens are one, and not only are they three-D, but they have holographic capabilities. Students do have to go to a classroom environment with other students and a teacher, who is a robot, for two consecutive days a week. In those two days the teacher monitors their progression. The teacher in the school environment checks what they have learned during the week in their home-schooling sessions. The teacher sets what the child learns in his home sessions the rest of the week, and provides the relevant data to his home computer. The home computer then teaches the student according to his teacher's instructions.

By the age of eleven, a child completes the first stage of his education. Their education differs vastly from our education. By this age, the minimum a student would have completed is the equivalent of a university degree. Subsequently, students continue their education up to the age of twenty-one. They don't learn in the same way that we learn; their advantage is that their brains are already developed at birth, and their subconscious minds are preloaded with knowledge. Additionally, their conscious minds are open to their subconscious minds. These factors mean, in simple terms, that they find it easy to learn, and that they can absorb comprehensive knowledge. Inevitably, children mature at a young age.

At the age of twenty-one not only are you fully qualified, but you are able to handle whatever life presents to you, which includes whatever technology is presented to you. You will have completed the equivalent of many university degrees. You will have specialized not in one field of study, but in all fields of study. Thereafter you work, because you are now fully equipped to handle work. Until you complete the entire educational curriculum and attain the compulsory academic qualification you are not, which means that educational qualifications are mandatory. On this basis, you have no choice but to complete all of your studies. As a part of your studies, you will have become fluent in all the languages of your neighboring civilizations, and those that are relevant to your civilization. This means that these people are likely to know many of our languages on Earth.

At twenty-one years of age you are not given the option to choose your career or job. You are given a job. Having said this, the conscious computer knows your mind better than you know it; it knows what suits you, what you want, and what will make you happy. Your success and happiness are important considerations. Having a job after graduating is compulsory. From then on you work, but one thing you will never do is stop learning, for, technology does not ever stand still; technology grows at a pace faster than we can imagine. This is why you have to upgrade your intelligence on a constant basis. Although you qualify in every field of specialty, and then are given work in one field of specialty, you have to keep up to date with every field of specialty. Learning is a process that continues for the term of your life, which aids in your evolution as a human.

Up until twenty-one years of age you live with your parents. You cannot leave home before this age because you are not ready to face society without the high standard of education that it takes to be a member of their society. Moreover, without your education, you simply cannot survive in their society.

When you are ready to leave home you are given your first unit to live in: a single bedroom one. This is given to you free. As we know, units are all the same, apart from the number of bedrooms they contain. The largest unit available has three bedrooms, and no more.

When you decide to marry and have children, you upgrade to another unit that can accommodate a family. A typical family comprises a male and a female, and two children. There is a two-child policy, which means that you cannot have more than two children. If you take into account the length of their lives, imagine how many children they could have and how big their families would be if they were given the freedom to have more than two children. There would never be enough Earths in the universe to house them! Even with the technology available, you do not choose the gender of your baby. This may occur if the society becomes unbalanced, in which case the governing authority will intervene in the matter. Fate, then, determines the sex of your baby.

There is the institution of marriage in their society. A typical marriage ceremony does not involve signing papers. There are no witnesses. There are no churches. Churches do not exist in an advanced, intelligent civilization; they do in a society driven by the negative side. The couple may have a celebration at home, at a venue, or even at a restaurant. One thing we should have learned from our society is that big events do not make for a happy marriage.

The official part of the ceremony involves an attending robot, whose job is similar to that of a marriage celebrant. He is what this book calls a "philosopher." The officiating philosopher weds a bride and groom in front of a gathering of their loved ones. I saw a ceremony on television, and it was similar to our wedding ceremonies, with several notable differences. I even said to Guyd, "We do the same thing."

First, the philosopher has the bride place her right hand on a table. Then, the groom places his right hand on top of hers. This is a symbol of their union. For half an hour they listen as the philosopher preaches his philosophical words to them. He explains what is expected of them and their marriage. He tells them what they should achieve in their short life – as he puts it. The following is only some of what he said, "Study is one priority of the family, which means that you have to be teachers to your kids. Study is also important for you both as long as you live, for, you have to be contributors to the society, and you can only do this by coming up with new ideas, and new inventions that are beneficial to the society. Whatever you achieve, I think you can achieve it together. You two are going to walk the path that we all have walked." The philosopher places a great deal of attention on learning, developing the brain, and making sure that the kids do the same thing.

When the philosopher finishes, he says, "Now you can kiss the bride." With that, the couple is married. A short ceremony follows. During the ceremony, the bride and groom cut their wedding cake. The wedding cake is sweet, and it symbolizes a wish of luck and happiness. They also open the dance floor as husband and wife. After a while, they leave for their honeymoon. The party may continue without them.

The bride I saw was wearing a white gown. "A bride wears pink in the instance that it is not her first marriage," Guyd said. I was a little surprised. "There are second marriages, but only because deaths do occur in our society. Accidents have resulted in people losing their lives. When these people marry, they marry for life." This was surprising to me, considering the term of their lives.

What we probably want to know is how these people age, and how they look when they reach an age of 1500 years. At 1500 years of age they look similar to a young-looking seventy-year-old. This tells us that you will see people with gray hair and wrinkles in their society. At that age, the type of body they have does show signs of wear, to the extent that many would wish to be recycled into a body that can function better than their aged one can. While this may sound like suicide, it is not. In our life span, terminating a life because it has aged or is not as functional as it could be is often regarded as committing suicide, and it is looked upon unfavorably, on many occasions with drastic consequences to the soul. Anyone in that society who wishes to recycle his body approaches the governing authority and presents his case.

A governor does not generally stand against such a decision made by an Atlantean. There are two options available to you as an Atlantean. The first involves dying a natural death and then following a process of reincarnation, in which case you are reborn on a different planet, in a different civilization. The second involves a process of recycling your body. The latter is interesting because it involves converting an old body into a younger body. The process takes three days to complete. What the process involves is having the person lie in a special cylinder, which is see-through. This book has termed this a "regeneration cylinder."

In the case of someone having died of old age (whose expressed wish was to live again in the same body with a continuation of his life), he is put through the same process in a regeneration cylinder. Once the body of a person is put into a regeneration cylinder, there are two specific stages. The first stage involves the emission of a pink fog. This has something to do with replenishing the body's blood. The next stage involves a blue fog. In this stage there is a process where the cells of the body are replaced. Three days after entering (or being put in) a regeneration cylinder, a technician opens the cylinder and an Atlantean steps out of it a younger version of himself, with the same personality, the same knowledge, and of course the same soul. The process has reversed his body clock, and he has become the equivalent of a five-hundred-year-old. Apart from having a younger body, he feels no changes. He will leave the cylinder and continue to live as he did previously. This means that if you choose it, death ceases to be a factor in your life. These humans can physically live for what may seem for ever, and have outgrown the need of the reincarnation cycle.

A person going through the regeneration process will then continue to be the same person, with the same life. Even though he has extended his life, he is not given the option to have more kids if he has already had two. Recycling in this way is an important decision because when one partner does it, the other partner will as well.

In some rare instances, someone may have a desire to start from scratch with a completely new life – that is, be born again in a process of reincarnation. However, this would have to be on another planet, and he would not be able to decide where. You will find that in the civilization of Atlantis, the souls that are reborn into babies do not come from an Atlantean that has passed away. They come from souls that in all probability have been upgraded from what is regarded as a lower civilization. This type of soul has obviously earned the right to be upgraded into an "up-market" civilization. Most Atlanteans prefer to continue their current life, the one in which they have lived for around 1500 years, and possibly even more, depending on how many prior cycles of recycling they have gone through.

Some people just want to die a natural death, because, when your life span is so long and you have recycled so many times, you can get bored with your life. Many have chosen to come to live with us on the surface because they want new challenges, and our lifestyle certainly offers them challenges. (More on this is discussed in another book.)

Let us take a hypothetical scenario: you as an Atlantean choose to die and then be reborn back into the Atlantean civilization. What would happen is that you would turn out to be the same person you were in your last reincarnation, before you died and were reborn. This would be noticeable after you turned ten or so years of age. This is because you have the same soul. Parents are pretty much the same on Atlantis. Education is the same. The parental and environmental factors cannot alter you from your past reincarnation cycle because there is little variability. Added to that, you can remember your past life almost from the outset, so your instincts are identical to those of your past life. Really, you are continuing your last life, just in a different body. In theory, you would shed your old ties and start with new ties. Yet this is not so simple, because your feelings will all come back to you, and you will most likely seek out your old partner and family. It would not be long after you are born that you would have a full recollection of your past life and of your past loved ones. Your natural inclination would be to seek them out and reinstate your past ties. This is why it would be logical for you not to die and be reborn in the same society, but to continue your life, just one thousand years younger.

For us intellectually encumbered humans, the reincarnation cycle works because we cannot recollect our past lives. Even so, we are still drawn to past partners, which comes to light as love-at-first-sight encounters. Perhaps we can imagine being in their shoes in light of this. This is why it would make no sense for these people to be reborn in the same society if they decided on the option of a natural death.

In view of the life span of these people, it would be logical to surmise that couples having children is only predominant in a society which is in a colonizing phase. This is the phase the Atlanteans in the colony in the Pacific Ocean are presently in. By contrast, in an established society there are only a few children. What you will find in an established society is that nobody wants to give up his existing family and have to start all over again by a process of reincarnation. Therefore, while there is a cycle of life and death, where people grow old and can die, they merely recycle themselves into a younger version of themselves.

13

The Governing Authority

The success and very existence of a society are reliant on its governing authority. In this world, our experience with governing representatives has been anything but ideal. Indicative of this is our snail-pace advancement over the ages. Then, we ask ourselves, what attributes should a governing authority have? Initially, the answer to this may not be so easy to digest. Plato got it right: long ago, he proposed that our leaders should be philosophers. We should be wary of our interpretation of the word philosopher. By this we can elaborate and say, true philosophers – humans who are not subject to the temptations of the negative side. To qualify for such a position, such an attribute would discount all humans, without a doubt. Then we may ask, who does that leave to govern us if we are to discount ourselves from the equation? Who could qualify as a true philosopher? While these may be thought-provoking questions, the answers to these questions may be hair-raising to some.

First, there is no such thing as a democracy when it comes to most superior societies. The majority of superior civilizations are governed by a theocracy, with theocracy defined as a governing authority in which a god is considered as the figurehead. Some of us should not jump ahead of ourselves at this point and start pounding our cloth-covered hairless chests; the rest of us should stop our alarm bells from ringing. This is not a theocracy, as we know it. We should not be shy but be frank and honest in admitting that we have had our fair share of so-called representatives of a deity running our societies, and our fair share of repression and decline, to put it diplomatically, under their authority. No, this book is not speaking of a return to ages of darkness, intellectual suppression, and outright deception!

In a theocracy, you as a citizen will never have the opportunity to make decisions on behalf of others. You will never have a true position of authority over others. You will be free to do everything. You will have everything an intelligent person could want. But . . . you do as you are told! In a way, there is no free speech, as we know it. You cannot promote a cause and then institute that cause in a society when it contravenes everything that a positive force stands for, but conforms to the philosophy of a negative force – to the extent that many of us lose touch with what is morally right. In a theocracy, the boundaries are defined and there is no opportunity to redefine or cross them. What is right and what is wrong determine those boundaries. Only in our ignorance can we challenge the boundaries of right and wrong. Never in such a society can we challenge these boundaries.

On our planet, not too far off in the future, we can expect to have this kind of governing structure instituted on our planet. Set boundaries of right and wrong will challenge our planet in an interesting way. For those of us who want to live in a "perfect" society – as perfect as one can be, given the challenges of a human – then this will be a welcome relief. The days of self-interested parties ruining lives will end. The days of those having too much and those having too little will also one day end. There will be many ends that go with this new beginning . . . to the dismay of some; to the delight of others.

Moreover, there may be much that is not good in this world, but there is still much that is good, and it is so easy to not see what is beyond the not good. Yet from thorny wild overgrowth can come the sweetest of fruits, and hiding in that thorny wild overgrowth can come from different seeds the sweetest of scents.

We must remember that an honest government makes for an honest society: a dishonest and corrupt government makes for a dishonest and corrupt society. The Ottoman Empire is a classic example, which proved this. When things are going wrong in a society, we must look to the very top. Australia, at the time of the writing of this book, is a case in point. The government sets a negative example, and that negativity seeps into all aspects of society. There are some fitting words of a wise man: The best can only come when a society is run by the best. This can happen when true philosophers rule us.

༄

Atlantis has several governors. The number of governors running a society can vary according to the population and the intellectual development of its people. The difference between a governor, who is a robot, and all the other robots in the society, is that the governors have a soul. The soul of a governor belongs to the original family of consciousnesses (first-generation consciousnesses) that came to exist in the universe – some may call these collective consciousnesses a god. (The details of the evolution of the universe and the creation of consciousness are described in *The First Cause, Volume I.*)

Every governor has his own office in the control center of the mother ship, and he is directly attached to the conscious computer. Apart from having a relationship with the computer, in a sense, the people have no personal relationship with the governors.

The governors are the brains behind the computer network of the mother ship. The governors have at their disposal many consciousnesses – the number can be in the millions, perhaps in the billions, perhaps even in the trillions (the amount can vary, and is determined only by how many they require) – that work with them; these consciousnesses, which are also known as "parts," do not have a physical form. How best to describe these parts is to imagine them as independent consciousnesses which have no physical presence, but which can interact with the physical presence. They communicate with the governor they belong to, and perform whatever duties they are given.

The parts of a governor interact with the computers and robots. All are telepathically linked. In this way, the citizens are able to communicate with a governor. Let us consider one application of this: a circumstance that a citizen of the society requires advice or help of some kind. Perhaps the citizen has become too old and is near death, and wishes to recycle his body into a younger one. The citizen will turn to his computer, which we know is directly linked to the governors through their parts. The computer will give the citizen appropriate advice. This means that there are no social workers or psychiatrists in a society such as this.

The governors tend to keep to themselves. When they interact with the population of the colony, it is merely as a matter of courtesy.

14

The Kitchen

I had so many things I wanted to do and see. I was particularly curious about the kitchen. Sure enough, every unit has a kitchen, just as every home in our society has one. While their kitchen functions in the same way that our kitchen does, in principle, their kitchen is to ours what ride-on mowers are to cattle as lawn-cutters.

You have the décor menu from which to choose your preferred décor. A typical design includes cupboards for storage, a bench, a sink, a fridge, and what at first glance may appear to be an oven. I saw cream-colored dishes that were as light as plastic. There is rarely any kind of food preparation, so there are no pots and pans. There are chopping boards in case you want to chop something, such as bread. There are also utensils at hand in drawers, and there is cutlery. There is even a unit that dispenses cutlery, which this book calls a "cutlery dispensing unit."

Displayed on the cutlery-dispensing unit are photos of cutlery items. When you select the unit of cutlery you want, it will appear in a service area of the unit. The technology used to reproduce items of cutlery is based on subatomic particle conversion, where subatomic particles are put together to create a matter form. During the process a red light displays. The cutlery-dispensing unit, then, is a specialized subatomic particle converter.

The kitchen has standard features and items; when any of those items are used and then disposed of in the sink, they are automatically replaced. Perhaps it could be a utensil from the drawer. Whatever comes in the standard design of the kitchen does not have to be purchased.

In the kitchen there is a fridge with double doors that slide open sideways when you either touch the fridge or think about opening it. While their fridge and our fridge have something in common in that they keep food cool, there is nothing common about their contents. Inside their fridge are tablets. You may rarely see a leftover pizza in one of their fridges, but you wouldn't eat that because you're better off getting a new one. There are no pets for you to feed leftovers to, either. They don't, by the way, have dogs in their units. Or cats. As mentioned in an earlier chapter, they do not have biological pets. They have cybernetic pets, which act like and have all the instincts of biological pets; these pets can even have an intellectual conversation with you. Yet, for all their instincts and similarities to their biological counterparts, robotic pets do not have a soul.

There are no dishwashing units in a kitchen of this kind, principally because there is no such thing as washing dishes. There is a "water" outlet along with a sink. As in the case of the bathroom, it does not emit water. It emits a blue fog. The sink is also a disposal unit. This means that if you have anything to dispose of you simply place it in the sink. This includes your cutlery, leftover food, and even plates. Once the sensors detect items in the sink, the tap will emit a blue fog that will turn everything in the sink into subatomic particles. If you have your hands in the sink, then they are cleaned, while the dishes are treated in the same way that dirt on your hands is treated. Foreign elements are converted into subatomic particles. The blue fog then disappears into the recycling system.

They have a television screen on the wall in the kitchen. It can serve as a three-D picture. While they can watch television shows, the television screen is at its best when it serves as a "window." So you may have on screen a waterfall, as I had in the bedroom. You may have a mountain scene. You may have a snow-laden landscape, or a field of tulips. Your personal taste or mood at the time dictates what scene you want. You may have a view of the ocean. On the other hand, you may have a stellar scene. Perhaps, even, the motion of the flying saucer when it is traveling in space. Perhaps a picture of the navigational laser locking system searching for and locking on to planets, when you are traveling in space. After all, this is not exclusive to the navigational center but available to everyone to view. (A description of the navigational laser locking system is provided in a later chapter.)

As we know, these are not static three-D pictures on a television screen; they are interactive, which means that the view on the screen produces all the relevant aromas and sounds of the scene. It even emits the temperature of the scene. If there is a resplendent sun, then its rays will stream into the room and add a mellow touch of warmth. The air you inhale will be the air that is emitted from the scene. A polluted scene will produce a grimy air, while a scene with a swathe of flowers will emit a redolence that will overwhelm your senses. The room you are in will become a part of that scene. If the scene is wintry, you may end up putting on warmer clothes. If the scene is summery, you may start undressing.

You can pick a theme. The screen will then give you a window effect, and you will choose what you want to view outside the "window." You may even see birds flying in that scene. Even butterflies. You may see a spider building a web outside your window. You may have a lake scene, with fish jumping up from the water; perhaps you will see birds feeding on those fish. You may have a waterfall scene, with red salmon fighting their way against the current, and defying all odds by jumping up the waterfall. You may also have bears feasting on their catch. You may have a cliff scene, with battering waves. You may have thunderstorms and a dazzling lightning show on your screen. You may be less inventive and decide to watch our Olympics. What you probably don't want to know is that they may even have a visual of you and your partner having an argument. They may find you entertaining. You and your life may be a reality-television show to them. They may have on screen any number of things that are occurring on this planet. The possibilities are endless. While this may be so, we must be aware that what they watch is censored. More on censorship is explained in a later chapter.

15
Food Tablets

It may surprise us to know that the people of Atlantis consume the same foods that we consume. They eat pizzas, bacon and eggs, fried chicken, fish and chips, and even hamburgers. They relish eating ice cream and all the desserts we know. They drink soft drinks – a healthy version, however. One way or another, many of our recipes have been handed down or inspired by them. Everything we think we are discovering is not a true discovery, but a rediscovery; we are reinventing or rediscovering something that has already been invented or discovered by others who were once in the same evolutional stage as us.

One thing you will not do in a kitchen of theirs is spend much time cooking! While we may share the same culinary tastes, there is a marked difference between their food and our food.

For a start, the food they eat is not fresh in its true form. Nothing comes straight from the paddock to the shelf. There is no such thing as fresh meat. While there are some fresh foods that you can purchase, such as fruit, most food does not come in this form. Much of what they eat comes in a tablet form. Food tablets are as light as a flake of snow. Once food tablets are converted from a tablet into an edible product, they become a replica of the original product. In a nutritional sense, there is no difference between eating a fresh steak and a steak converted from a tablet. If anything, their food would taste better and nutritionally be better.

Anyone who adheres to the vegetarian philosophy of not eating meat because it has been sourced from a living being should take note. Non-intelligent animals (determining what is considered intelligent will be covered in a later chapter) are meant to be a part of our diets just as the plant species is. All plants are as conscious as animals are, with a metaphysical component forming the basis of that consciousness.

Now, what may seem a challenge to understand is that there is only one so-called cooking apparatus in a kitchen. We may at first glance assume that it is an oven. In some respects, it does have some vague comparison to an oven. But, don't be fooled, this is no ordinary device. The appropriate title for this is a "food converter."

As stated earlier, their food comes in a tablet form. Let us say that you want to eat a pizza. You have a refrigerator in your kitchen in which your food tablets are stored. You can identify what meal a tablet is by its front label. On the packaging of a tablet, there is not only a picture of the product in its cooked state, but all the information necessary for you to know about the product. The tablet is about the size of a twenty-cent Australian coin. You take out a pizza tablet and put it in the food converter. You do not have to set a timer or a temperature. The converter reads the chip on the label of an item, and then it does all the work. You do not even take the labeling or packaging off your tablet before you place it in the converter. On the contrary, you must leave it on so that the converter can read the mathematical signature of the food item. In no time, the food converter will convert the pizza tablet into a ready-made pizza, with all the trimmings, according to the flavor you chose. It then looks and tastes exactly like a pizza.

Water also comes in a tablet form. Let us say you want a cup of coffee. You put the cup of coffee tablet into the food converter. Different types of coffee tablets are available: some with one sugar, some with two sugars, some sugarless, some black, and some with milk. Accordingly, you buy the coffee to your taste already prepared in a tablet form. You don't even have to put your tablet in a coffee mug, because the mug is a part of the tablet. Everything is considered; for instance, it is not too hot that you cannot hold the mug. Only the contents of the mug are boiling hot. If you order your coffee sweet, you will not even have to stir it. All you have to do is put the coffee tablet in the food converter, and not long after you will have a mug of coffee ready to drink.

The meal tablets even include a plate. For instance, a bacon and eggs tablet will go in the converter and come out looking like a regular bacon and egg meal, on a plate from which you can eat. All that is missing is the cutlery, but a different device provides this: the cutlery-dispensing unit.

In the case of a pizza, it will come out of the converter on a serving plate, already carved. However, you do have a chopping board in your cupboard and carving knives in your drawer that you can use if you ever have a need for them.

What many will find just as interesting is that when you are finished eating, you put everything – the scraps, plate, and cutlery, along with the chopping board and carving knife if applicable – into the kitchen sink. The tap will sense the presence of these items and subsequently emit a blue fog that will convert them into subatomic particles; the blue fog will then go through the recycling system.

If you used the chopping board and the carving knife, you may wonder what then happens to them: do you get replacement ones? The answer is yes. However, we should note that items are replaced when they are a feature of the standard kitchen. Standard kitchen refers to the kitchen as it appears in the décor menu. This means that anything that is not a standard item of the kitchen must be purchased, such as food. Your chopping board will always reappear in the cupboard because it is a part of the standard kitchen. This applies to the carving knife. When you make any of these items dirty and throw them in the sink, they are gone. Once these kitchen items are broken down into subatomic particles, the nanorobot will know what was broken down. So if an item broken down was something from the standard kitchen, for instance the chopping board, it will automatically be replaced. Indeed, anything that is a feature of the standard kitchen is automatically replaced once it is used and disposed of.

In this kitchen, you cannot drink water from a tap. Water is purchased in a tablet form. Some things are available to you free because they belong to the standard design. You have to purchase everything else, which is what maintains an economy. No society can survive without an economy. Without an economy, people will not work; a society cannot be productive when its citizens do not work. It is not advisable to have an unproductive society; hence, an economy is imperative. The consequences of not having an economy are far-reaching; they are capable of leading to the downfall of civilized man, as he exists in a superior society.

There is another way of preparing your food. All you have to do is think about what you want to eat. Moments later, you go into the dining room and whatever you thought of will be there ready for you to eat. If you were out and about to come home, and you were feeling tired and didn't want to waste time cooking (that is, taking a tablet out of the fridge and putting it in the food converter), then you can have this done for you; it is the computer system in your unit that will do it all for you. This means that if you order fried chicken, then the fried chicken tablet will hover into the food converter. When it is converted into your meal, the fried chicken on a plate will hover out of the converter to your dining table. This won't happen in slow motion; you won't be able to watch it drift leisurely past you, nor will you be able to knock it over. Indeed, it will simply pass through you. It just seems to happen in a flash. Your drink tablet will have gone into the converter when the fried chicken tablet did, and it will also hover with the plate of fried chicken to the dining table.

Although you see the process occurring, the items going through this process are almost see-through, or fog-like. When the product rests on the dining table, it instantly goes from that state to a regular state. The fog-like state is some form of energy state that can pass through matter. Nanorobots have the technology to exist in this energy state and make other forms of matter exist in this energy state.

The required table settings will be set on the table for you in this instance. Table settings are a part of the standard design of your dining room. The computer will assign the appropriate table settings whenever you need them. When you finish eating, you won't even have to clear the table. The computer can clear it for you; it will clear your plates with meal scraps on them. Everything is converted into subatomic particles. However, as mentioned earlier, women usually take delight in kitchen chores. They clear away the dishes and deposit them in the kitchen sink.

It almost seems as though you have a personal servant; however, there are no servants in your home. If there is a special circumstance that you do need one, one may be allocated to you free of charge – male or female, depending on your request. On occasions, you may be entertaining a large group of guests. In this case you can hire a servant just for the occasion.

Let us consider a different circumstance: you crave a pizza for dinner but you do not have a pizza tablet left in your refrigerator. All you would then do is ask your computer to order one for you. You are asked by your computer if you want it cooked or uncooked (in tablet form). You make a choice. Everyone prefers to have it delivered cooked. Why would you want to waste time "cooking" it yourself? Frankly, just as our pizza from the supermarket is not as good as the one from the local Italian pizza store, so with their tablets from the supermarket. It won't take long before a delivery vessel arrives with your pizza.

The delivery vessel is in the shape of a shopping trolley. It will park on the road outside your balcony. If a robot delivered it, then he will inform your computer that your pizza has arrived. Your computer will then notify you, telepathically, that the delivery has arrived. The robot in the delivery vessel will at all times converse with your home computer. Your home computer will confirm that you ordered the pizza, and that you will be right out to collect it. You usually step out onto your balcony to accept the pizza. The pizza will most likely hover to your hands.

Naturally, you have to pay for the goods you ordered. You usually don't pay by exchanging money or cards, but by putting your hand on a glass panel on the delivery vessel. The panel is flat, square, and horizontal. A green light scans your hand, and the information registered automatically goes into the main computer system, which transfers the money to the shopkeeper. Your identity is registered from your handprint, which acts like a credit card.

If you decide you do not want to pay using your handprint, you can pay by cash, or you can put the outstanding amount on your account.

This tells us that an advanced society still has banks. Banks are private enterprises; however . . . these banks are not motivated by looking after the interests of the shareholder. There is a completely different culture, philosophy, as well as mindset in the banking industry in their society.

When you thank the robot for delivering your pizza, he usually responds, "It's a pleasure."

We may be wondering about the physics of the food tablet – how a tablet can convert into a cooked meal. The answer to this can be found in Book II, Chapter 7. Challenge your brain and think about it; see if you can come up with the answer for yourself! I must admit, it had me perplexed for a spell – the simple things often do.

16

The Balcony / Garage

Following the footsteps of Guyd, I stepped out onto the balcony. We know that there are six floors to the living center of the mother ship. On your floor, someone probably lives in living quarters on top of you, underneath you, and beside you. Living quarters have ceiling heights that are higher than our normal ceiling heights. All living quarters have a balcony. A special glass-like structure frames the balcony, which is not very high. Without precautions, this could pose a threat: it is possible for someone to fall over it accidentally. Yet in a society such as theirs, all possible safety precautions are considered.

For instance, you may have an adventurous toddler who climbs up onto your outdoor furniture and then accidentally falls over the edge of the balcony. This is a likely scenario, and it has occurred. What happens to a toddler when he falls over the edge is that he will just hover next to the balcony until someone discovers his predicament. He cannot move from the spot. In this instance, a robot will be the first to know. The conscious computer will inform him – it knows everything that goes on. Within seconds, a robot will be at your balcony, and he will put your toddler back on the balcony. You as a parent will not even know what has occurred to your child. The robot will not tell you, either. He does not want to upset you and encourage any negative emotions in you. The objective of a human is to stay positive at all costs.

Just outside every balcony, there is an antigravity force in place. This preventative measure will not permit you to fall to your death if you fall over the edge. We could do with such technology in our society!

When you look over the edge of your balcony, you can see a drop that extends all the way down to the ground floor. You may even see people walking down below, depending on from which height you are looking. You can have greenery growing on your balcony, such as pot plants and small flowerbeds. There is a special light built into the ceiling to allow for the growth of greenery. The balcony is olive-gray in color. The neutral color scheme seems to be applied everywhere.

On the balcony, you are likely to have an outdoor table and chair set. You can sit there and enjoy the fresh air – there is no such thing as pollution – and you can watch everything going on around you on your street, as well as below and even above you. The view is impressive. From there you can see all of the other units; apart from some minor personal preferences, all the units have the same features.

Beyond your balcony, the street is about two hundred meters wide. Remember, this is not a street with a surface; it is an area of space that extends from the bottom floor to the top of the sixth floor of the mother ship.

Next to every balcony, there is a garage. There is no balcony before the garage. The garage is not immediately obvious; any suggestion of a garage is a horizontal line in the middle of a wall. This is actually a door that slides open in an up and down motion. You cannot have two or three flying saucers in your garage. Each unit is limited to one. To have an idea of the size of a garage, we need to imagine a four-car garage.

You have space on your balcony for your shopping trolley. We would not immediately picture how this shopping trolley looks. Nonetheless, this stays on the balcony. Shopping trolleys vary, in that some are equipped to seat children. A shopping trolley is collapsible to about the size of a briefcase, and it is made of lightweight material.

The best way to picture a shopping trolley is to imagine a rowing boat that seats two adults side by side. There are no doors, as in a boat, and there is no roof. It generally hovers around two feet off the ground. Your toddlers are restrained in seats behind you. These seats are designed for toddlers who are around six years of age or younger. If you have older children, you usually have a second shopping trolley. In addition, if you are shopping, and if one is not big enough, another trolley can be "towed" about eight inches behind your first one; this is invisibly attached by a magnetic force.

At the front of the shopping trolley is an area allocated as a storage area. This storage area occupies approximately one-third of the shopping trolley. It is temperature-controlled for goods that need to be kept refrigerated, such as food tablets. The shopping trolley does not need to be large because all of the groceries are in a tablet form. Furthermore, anything that does not fit into your shopping trolley will be delivered to you.

The shopping trolley is connected to the computer network of the mother ship. It has antigravity technology, which allows it to hover. You don't push it; you just tell it where to go. It can even operate without the presence of a human.

The shopping trolley belongs to you, and it recognizes you. This means that no one else can access it. It knows all your identification details, such as where you live and all the codes required to access your quarters. There is no such thing as identity theft in a civilization of this kind. Furthermore, anything you purchase is safe in the trolley in that no one else can gain access to anything inside it.

You can request your shopping trolley at any time, just as you can send it to whatever location you want. For instance, let us imagine that you have just done some shopping and your trolley is full of goods. You then decide to go somewhere directly from the shopping centre. Instead of taking your shopping home and unpacking your goods, you can instruct your shopping trolley to do these things. According to your instructions, it will go to your quarters and instruct the goods to unload. You do not need to be a part of this process. When you return home, all of your shopping goods will be in their appropriate place of storage. Food tablets, for instance, will be in the fridge. Clothing will be in the wardrobe. The computer in your unit will coordinate all the maneuvers for this to happen.

17

The Main Shopping Center / A Small Boutique / The Physics of Clothing and Shoe Adjustment

I was looking forward to seeing how the shopping trolleys work in a shopping center environment. Even though there are facilities for people to shop from their home computer, there will always be facilities for people who want to physically see goods and feel them in their hands. Shops will always exist, one way or another – particularly specialty shops.

Guyd requested the lift to appear, and we entered into it. He then told the computer to take us to the sixth floor. He had to provide the computer with a code, and within around twenty seconds we were there. The lift itself had the usual olive-gray color scheme, with 24-carat gold-plated handles on the walls. I noticed that the lift made no noise, and I never once felt the motion of the lift. The entire process was quick. We just stepped in, and in less than a minute the doors opened to indicate that we were at our destination.

Guyd told me that you are not meant to take bulky items such as your shopping into a lift – after all, your shopping can be delivered to your unit by a shopping trolley. Guyd also told me that they do not have escalators, that they only have lifts.

The lift opened to the heart of the shopping center. The sixth floor is strictly a shopping precinct. Before my eyes was the shopping center of the future. People were everywhere: some were walking, while others were hovering past in their shopping trolleys. Outdoor cafés were bustling with people. People were eating, drinking coffee, and enjoying themselves. I nearly passed out when I saw how magnificent shop fronts are.

Guyd said, in response to my astonishment, "They are always illuminated." I saw dazzling lights. I couldn't take my eyes off the diamond walls and diamond features. I cannot convey how welcoming the place was to me. What is interesting is that the shop fronts are not vertical. They have a 45-degree angle that slopes inward. The angle extends up to the roof and serves a purpose, of which we will learn.

When you go shopping you don't usually take your flying saucer with you. This is why you have a shopping trolley. You may decide to take the lift to the sixth floor, and then stroll to the shop you want to go to if it is within walking distance. If it is further, you are likely to take your shopping trolley. If you are already at the shopping center and then decide you need your shopping trolley, all you have to do is telepathically request it and it will automatically leave your balcony and find its way to you. You have to give your shopping trolley the code that represents your location, or where you want it to meet you.

If the retail outlet in which you are planning to shop has a policy of allowing your shopping trolley into it, then you can take it into the store with you. Most stores don't welcome your shopping trolley for several reasons, one being that you can jam up the store if many customers come in all at once. By having you use the store's trolley, the store controls the traffic flow into the store at any given moment. Most stores insist that you leave your shopping trolley parked outside of the shop.

Every store has its own shopping trolley to suit its store and its products, and these come in different sizes. If one of its products doesn't fit into your trolley, the store will arrange to have it delivered to you. Sometimes the store's shopping trolleys are much smaller than the regular-sized one everyone possesses, which makes it easier to maneuver through the aisles. When you enter a store, you tell the robot at the counter what size suits you, to cater to your family size.

Now we can explain the reason that the shop front is angled 45-degrees. The space created by the shop front's angle is an area for your shopping trolley to park and wait for you. By park, this means to hover. Some shops only have a small shop front, which only enables several shopping trolleys to park there. Other shops are large and have supplementary parking spaces allocated for shopping trolleys.

Your trolley will not park on the street. It will let you out in front of the shop. Then it will find a parking spot either in the angled space above the shop front or in a specific trolley parking-bay. Most shops know their turnover of people, how long those people will stay, and what they will be buying; therefore, they usually have the required parking spaces to cater to the projected amount of customers visiting their store. In some respects, it is similar to our street shops, in that there is only a limited number of spaces in front of shops, which means that the shopper needs to find a larger car park nearby in which to park. In the shopping center on the mother ship, there are parking bays that have space for thousands of shopping trolleys. When you finish shopping in a store and require your shopping trolley again, you merely instruct it to return to you. You don't have to find it, as it will find you.

As you shop, the goods you select are loaded onto the store's shopping trolley. When you leave the store at checkout, it will arrange to have your purchased goods transferred into your trolley. First, at checkout, you will exit the store's trolley. Then, while you are paying for the goods, the trolley loaded with your purchases will be in the loading dock transferring those purchases into your trolley. Robots will assist with the transfer. Finally, your trolley will meet you at the exit of the shop once you have paid for the goods. The process is a quick one.

In some circumstances, there may be a problem loading your purchases onto your shopping trolley. In other circumstances, you may be inclined to have your purchases sent directly to your home. It may be easier, then, to have all or some of your purchases sent directly there, and this is usually done by transmitting the blueprints of your purchases in the form of signals.

Every unit has a "converter menu," which is holographic. The converter menu is a log of items that have been purchased by you and transmitted to your unit. The converter menu is not static; it is able to appear anywhere you want it to appear in your unit. This means that when you do your shopping, you can have the retailer send the subatomic particle blueprints of your purchases to it. When you return home, you can request the converter menu to appear. On it, there will be a list of all of the purchases that you have made, which have been sent to you in their information form. Every product has an item number. You then scroll through the list and select the item number of the product you want converted into its matter form. You also instruct the converter menu of the location that you want the converted product to go to. The robotic technology associated with the converter menu reads the blueprint and creates the product from subatomic particles.

For instance, let us say that you were out shopping that day and purchased a large pot plant, which you want placed on your balcony. Let us also say that you have just arrived home and are relaxing in your lounge room. You request the converter menu to appear. It will appear in front of you. You browse through the log of items and select the pot plant. You quote its item number, and say that you want it converted and placed on the balcony, in the right hand corner, next to the chairs and table. In next to no time, the pot plant will appear at the location you have specified. The item will no longer be displayed on the converter menu. You go through the same process with all the other items listed in the converter menu. Often you will want to observe the placement of the item to be sure you are happy with its placement. In this case, you will tell the computer that you will be out right away to confirm that it looks appropriate in that position. If you don't like where it has been positioned, then you merely tell the computer that such a placement is not appropriate and that you want it placed elsewhere, for instance in the other corner of the balcony. The pot plant will disappear and appear in the new location in a matter of seconds. You can keep doing this until you are happy. Any time you feel that something is not where it should be, then it is up to you to tell the computer where you want it to be, and the computer will take care of moving it.

Before the goods are sent from the shop to your unit, you have to provide specific delivery instructions. For instance, you have to tell the shopkeeper to send the goods to your converter menu. The item will then go into a log in that menu, in a similar way that emails collect in an inbox. You refer to the converter menu in terms of a code. For instance, the code for the converter menu may be T45. You say to the shopkeeper, "I want this one to go to T45." P63 may be the code of your wardrobe particle converter, so if any purchases are items of clothing, you may say that you want the clothing to be delivered to P63. You may just send everything to the converter menu, and then afterward direct them to their appropriate locations.

One thing that this tells us is that these people have many codes to remember. However, they don't have a problem of ever forgetting their codes. Once a code is stored in their memory, they are capable of retrieving it at any time. Their brains are like computers, and they have photographic memories.

Shopping is an experience in itself. One thing about the shopping experience is that no one walks through a large store on foot. On a rare occasion someone might, but in most stores this is neither practical nor convenient.

The shopping trolley will drive you throughout the store complex. All shopping trolleys use antigravity technology, which means that they hover. Because the shopping trolley is telepathically tuned in on your mind, you can telepathically dictate where it should take you; alternatively, it will automatically take you to those products you want to purchase. If you want to look at a product that is three floors up on a shelf, then the shopping trolley will rise to the height of the product.

Once you see a product you want, you won't have to physically take it off the shelf. Every product has a chip on it, which is similar in principle to our barcodes. The shopping trolley reads your mind on which item you want. It will send a message to the chip on the retail product to hover off the shelf and into your shopping trolley. If by chance you want to look at the item up close, the item will hover to you, and you will then take it in your hand and look at it. Then when you release it from your hand it will hover away, either back on the shelf or into your shopping trolley. Where it goes is determined by your thoughts.

There are no shelf stackers or stock auditors in a store of the future. When most of a product on a tray is low of stock or empty, then at some stage when the store is quiet, a robot who is responsible for restocking the shelves will replace trays with fully stocked trays. He will bring out an industrial trolley that is stocked with the appropriate trays. Half of the trolley has trays of new supplies. The trays stack in a similar way that bread crates stack. The other half of the trolley is empty.

The robot performs no function other than to organize. Intelligent chips on the trays follow the instructions of the robot. Once a trolley is positioned right next to the shelf needing restocking, a tray on the shelf that bears a product, whether empty or low in stock, lifts and vacates its space on the shelf. It slowly hovers to the empty half of the trolley. These empty or near empty trays then stack up in that section. At the same time that a tray vacates a shelf, a full tray of the same item hovers to the shelf and occupies its place. You can see the trays – a vacating one and a replacing one – pass each other in midair in slow motion.

Every tray carries the one unit of product. The chip on the product is aligned to the chip on the tray, which is aligned to the chip on the shelf. Therefore, the replacement tray knows where to go. One replacement tray leaves the trolley at a time. Once the process is complete, the robot returns to the warehouse, leaves the trolley to another robot, and if conditions permit, returns with a different load of supplies that need to be stocked on the shelves.

Nothing rolls on the ground or is pushed or pulled. These trolleys use antigravity technology. The robot telepathically communicates to the trolley. It appears that central to their technology is the telepathic frequency. In our society, we are in the early stages of voice-recognition technology. Their society has the technology to recognize the telepathic frequency.

In the warehouse, there are robots who carry out different duties. One will restock a trolley. The trolleys loaded with stock will then be stored in the warehouse until such time they are required. The robot responsible for making sure that shelves are stocked never has to wait for a trolley. Trays are always prepared and stacked long in advance of them running low on the shelves.

The chip on the tray records the quantity of items on it. A robot in the warehouse monitors this record. The warehouse is aware of which stock is low and in need of replacing. The chips are all networked to a computer, and that computer is linked to all the robots. Everything is precise and organized. No one has to physically go to a shelf and take count or do stocktaking.

As far as the trays are concerned, the size is subject to the product. On the shelf that stocks the tray is a label. On that label is a picture along with the name of the goods. There is also a description of the goods.

*

Shops remain open twenty-four hours a day, which means that robots work in shifts around the clock. The greatest advantage in having robots in place of humans is that they do everything professionally and precisely.

Private individuals do not own the shopping center on the sixth floor; it is a part of the mother ship. Yet private individuals do lease shops. Frankly, no one owns property. Businesses can only take out a lease on a property.

While humans may own and run businesses, the majority of staff are robots. Indeed, robots perform most jobs. A human's participation in a business often lies in such things as the design and innovation of the products produced; setting up and overseeing the manufacturing of products; and organizing the duties of the robots. In no circumstances do humans have jobs that involve manual labor. Once a product is on the shelf for sale, humans just make sure that the manufacturing of products is to standard and current.

The individuals who have retail outlets or industries lease robots. The robots then work according to the duties that are allocated to them.

At some stage of my examination of the retail sector, I went into a small clothing boutique. I saw clothes hanging on racks; the technology was similar to that found in the wardrobe. Even though clothing and shoes (with the exception of children's wear) only come in the one size, people still want to try on clothes, and they do this in changing rooms. To ensure that the customer is comfortable with the service, a female will have a female sales assistant; a male will have a male sales assistant. People still feel a sense of modesty and discomfort with someone of the opposite sex staring at them, even if that someone is a robot. After all, robots have feelings – perhaps those feelings are not one hundred percent the same as the feelings humans have; still, they are close.

In their society, there are no clothing alteration stores. When you buy a pair of jeans, you will never have to adjust the hemline. There is never a problem of clothing or shoes not fitting you, irrespective of your body shape. Just how clothing and shoes are able to mold to your body is interesting.

First, it is in the production process that the material is altered by way of a chemical additive. No matter what kind of chemical process may be involved, logic would dictate that on its own the material could not perform the task of adjusting to the body, and that there must be some form of intelligence in the material itself additional to the chemical additive. This is because the actions of the clothing in terms of self-adjustment are random, varied, and not standard. The actions vary according to each individual, and are determined by body shape, height, and so forth.

Whenever there is a question of intelligence, one can usually translate this intelligence into a robot. When it comes to clothing self-adjustment, one can consider the involvement of a nanorobot. Essentially, in this instance a nanorobot has the job of a tailor.

What this tells us is that clothes have a nanorobot built within them. Until a garment is sold, the nanorobot is programmed to adjust the clothing to fit the wearer. If the potential buyer does not purchase the garment, the nanorobot will continue to perform this action as long as a potential buyer samples the garment.

The nanorobot has telepathic capabilities, and it reads the mind of the buyer when he is trying the garment on. For instance, if the garment is a pair of trousers, a female who is in the market for a husband may want a tighter fit, in which case the nanorobot will adjust the garment to suit the buyer's taste. A married woman will probably prefer a fit that is not so body hugging. In this case, the nanorobot will adjust the trousers to the tastes of this buyer, and the nanorobot knows when you are satisfied with the fit. This suggests that one trouser may be adjusted many times, according to the tastes of each consumer.

On its own, the nanorobot does not achieve this objective. As stated earlier, there is a dual process involved. The chemical additive is the second aspect of this process. The chemical additive in the material makes it possible for the nanorobot to adjust the clothing.

Now, once a fit is made, the customer may decide to buy the trousers. In this case, the trousers will not adjust again. Clothes adjust once to your body; you have to approve the final fit. Once you decide to purchase the item, and make the final approval, the clothing will never alter again because the nanorobot will have completed its program. Its one action is to adjust clothing to suit the wearer until a sale is made. Once the buyer is happy with the fit, the nanorobot's program has a termination setting. It is upon purchase, when the item is registered as sold, that the nanorobot's termination setting is activated. Additionally, the adjusting properties of the clothing, based on the chemical additive, expire. The first time the item of clothing is broken down into its subatomic particle state is the nanorobot broken down into subatomic particles.

If you eat too much for lunch and the fit ends up being too tight, bad luck! The clothing will not alter to suit your bloated size. Others generally cannot borrow your clothes, either, unless they are your exact size. It would not be good for business if people shared clothing, or if clothes constantly adjusted to fit your changing shape. There is the economy to consider.

One thing about the future is that there will be no post offices in the retail sector. Technology will see that there is no place in their society for the postman. In every study, there is a particle converter. This will replace the post office.

Let us imagine that you have a parcel that you want to send to a family member somewhere in the universe. What you can do is simply place a label on the front of the packaged parcel, which outlines the destination address along with the sender's address. Every address has its own code. To post an item, all you have to do is place it in the converter. What the converter does is it emits a green light as it scans the parcel. In the scanning process the address is identified. The converter converts the item into subatomic particles. The information on the subatomic particle blueprint of the item is converted into a signal. This signal is then sent to the address specified on the label. The signal is capable of being sent to the farthest reaches of the universe.

The parcel will then wind up in a menu, which is the equivalent of an email inbox, at the destination address. The signals are collected until you "open" them – that is, convert the signals into their matter form.

18

Robots

I only knew that there were robots as opposed to humans working in the retail sector because Guyd had told me this in advance. To look at them you would never know; they talk and look like humans; they even have half of our feelings – positive feelings. Despite not having negative feelings, they are programmed to defend themselves. For instance, let us say that you have the urge to do something detrimental to a robot. He can read your mind and know your feelings and intentions toward him, in advance. As a precaution, he will take the initiative and emit some type of ray. This has a calming effect on you, and it takes you out of your negative state. If it is a serious case, the robot will call a technician to take you to the equivalent of a hospital, which is actually the technician's workshop. Here, a technician will examine you to determine what is wrong with your brain: what caused you to shift to the negative side. Depending on the circumstances, your mind will then be "tampered with," and your intelligence will be upgraded so that you become a useful member of society and a good citizen. This process will have balanced out your negative and positive sides, so that your positive side predominates and your negative side abates.

Robots enjoy almost all of the privileges that humans enjoy. For instance, a robot has a job that is allocated to him, and he can work for as little as four hours a day, often on a shift basis. This means that he earns a wage to survive. As you would expect, a robot needs a wage not just to buy clothes and other essentials, but for entertainment and leisure activities. Some robots are involved in research, particularly in the field of robotics; others are engineers. Just as some have major responsibilities – such as being the pilots of the mother ship – others have duties in hospitality, retail, and customer service. Some provide the labor necessary in the manufacturing sector.

Robots are a functioning part of the mother ship and colony. The conscious computer has an officer (a robot) that is in charge of robot personnel, which means that he is responsible for the wages of a robot. Companies requiring robots lease them out from the personnel officer.

Robots are well looked after in society, and are maintained by technicians. Technicians are linked to the minds of all robots, so if anything goes wrong with a robot, a technician will immediately know which robot is in trouble. Every robot has an identification code, which technicians are linked to.

We may wonder about the life cycle of a robot. In answer to this, robots have much in common with humans. We know they sweat and stink, which means they must shower and groom. They also possess their own units. Sometimes three robots will share a unit, since the maximum number of bedrooms in a unit is three. Those humans with dirty minds who are reading something into this should note that robots do not engage in sexual activities. Robots enjoy leisure activities just as humans do, so they often go to the beach and read a book. They often go shopping. They visit all the recreation areas; they go fishing; they go on picnics. Some have their own research projects. They may enjoy writing, or they may be artists. Some are singers. In many respects, there is little difference between a robot and a human, but what difference there is, that difference is vast.

As indicated, robots do not engage in sexual activities. We must remember that the primal instinct and drive of sex exist for one purpose: to ensure the perpetuation of the species. Sex has not been provided for purposes of pleasure just for the sake of pleasure. There is a catch to sex: offspring. That the activity of sex is interlinked with pleasure ensures the perpetuation of the species. Robots are not a species that reproduce. They are created. Consequently, they do not engage in sexual activities, and they have no mechanism for reproduction. They may look like a human in every way, and they may eat, drink, and pass waste just as a human does, but the reproductive aspect of a human is preserved for the human. Sex and marriage are sacred, and occur between a male and a female human. No deviation from this model is permitted in an advanced society. This means that robots do not get married. Humans have an instinct to seek out a partner to fulfill their needs, and those needs are directly related to feelings that are attributed to the perpetuation of the species. In other words, humans have the primal instinct to seek out a mate to fulfill the sex drive, which in turn achieves the fundamental blueprint of sex: reproduction. Each sex needs the other sex to satisfy the primal instinct. On this basis, robots have no need of marriage. Neither sex nor marriage is a part of their programming. They have no means of reproducing, which means they have no need of, let alone any desire for, sex, which means they have no need of seeking out a mate.

We may find the whole subject of robot interaction in a human society a fascinating one. Many futurists are troubled by the concept of having robots in a society, and that sense of trouble arises because of the potential for the robot race to destroy the human race. This concern is a valid one, since a human can never match the intellectual capability of a robot. A robot has absolute intelligence with far-reaching capabilities. The potential for robots to take over and destroy the human race is likely if not for one barrier, which this book will clarify shortly.

What is poignant is that robots are meant to serve humans, not the inverse. Humans are the architects of robots. Humans have always been, and always will be, at the top of the ladder or food chain. Man, as opposed to cybernetic man, will and must always be superior to all other species of life. Nevertheless, he has to be vigilant because he can lose his placement if he is negligent or careless, and does not consider all of the consequences of any of his actions.

Let us return to the question of robots and sex. We know that robots are not involved in sexual activities. We know that they do not form attachments with the opposite sex or marry. This tells us that they do not possess the feeling of love. If robots were programmed with the feeling of love, the potential would always exist for complications in relationships involving robots and humans. The prospect of a robot falling in love with a human has far-reaching implications. Moreover, when you possess the feeling of love, boundaries cease to exist to prevent people from each of the two races from falling in love. Despite such a relationship being inappropriate, it cannot work, and it would lead to the decline of a society.

There are other significant reasons that a robot has limited feelings. Of the feelings he has, he can feel happy just as he can feel sad. For instance, Guyd was genuinely sad to see me leave when I did. However, that sadness is not the type of feeling that can lead to depression or suicide. The negative force that exists within humans, which does not exist within robots, causes depression. Accordingly, a robot can never feel depressed. He can never commit suicide. This also means he can never become violent or angry. He can never experience feelings of aggression. This is important because feelings of aggression and lack of control are ingredients in the recipe of disaster. He also cannot feel jealousy. Jealousy is a negative ingredient that can lead to extreme negative behavior and emotions.

This takes us back to the question of love. To possess the feeling of love is to possess the feeling of hate, as love and hate are two sides of one package, as hard to believe as this may be. One half of this package cannot be present in a human without the other half. They are the same feeling, only they are "aligned" differently. They are the one force with two poles: love is the positive pole; hate is the negative pole. This is significant because in the package of feelings of love and hate is the feeling of greed. You cannot possess the feeling of hate without possessing the feeling of greed. When a robot is in possession of feelings of love, hate, greed, and jealousy, then he becomes subject to circumstances that even we as humans cannot control; these circumstances will guarantee that the robot race takes over the human race.

Therefore, the barrier that prevents this from ever happening is to limit the feelings of a robot. The robot as he exists now does not have any negative feelings within him. He does not question the human master, whom he serves. It is not within a robot's program to question his master. This means that there is no possibility of a robot ever wanting to change his station in life; there is no possibility of a robot ever feeling unhappy in his role. He is only happy because he does not possess the negative side, or the negative force within him that would make him question his role and his happiness.

The subject of the negative force is a deep subject, and its history and relevance to a human are discussed in another book, which has already been mentioned elsewhere.

Perhaps now we know why a theocracy runs this type of society. If there were a human in place of a theocracy, there is no doubt that he would change the equation – that is, he would alter the status of a robot for self-serving interests. So long as man has a negative side within him, and so long as man has a position of power, he is vulnerable to corruption, and it matters not how intelligent or how evolved he is. This is why a human cannot have a position of power, unless he has a high capacity to reason. There are civilizations in the universe that are superior to Atlantis, and they do not have a theocracy of the kind mentioned. Having said this, they have their share of problems; their share of wars . . . of them, you would not want to know!

19

The Command and Navigational Centers / The Physics of Space Travel I

I was invited to visit the navigational center. This was the only area of the control center to which I was permitted entry. As the navigational center is technologically designed, there is no room for a human to obtain employment in it. This also means that, apart from the governors, only robots with a job in the control center have clearance to enter this area of the mother ship.

Even though I was permitted entry into the navigational center, I was not permitted entry into the command center. I was told that the command center is only concerned with combat and security. It is from the command center that they are able to protect the mother ship.

Guyd said, "We are always ready for combat, so that nothing can take us by surprise. Our policy is to avoid conflict, wherever possible, because we are not a violent people, and we will never initiate violence unless we are under threat, and then that would only be a defensive maneuver. This is why our spaceships have a cloaking system, which makes us invisible to attackers – physically and technologically. You cannot fight an enemy if you cannot see him or detect him with technology. If circumstances arise that we have no choice, we will blow the enemy craft up. No one stands a chance against us, which makes our technology a target. Everyone would wish for our technology. No one in the universe on the negative side has the power to defeat us in combat. No one on the negative side has the equivalent technology. There is a disparity in the technology of 'negative' civilizations and 'positive' civilizations. There are civilizations in the universe that are far superior to our civilization, but once a civilization is in this league, it is usually run by a theocracy. Without a theocracy, the likelihood of a civilization turning to the negative side is possible and probable.

"In our home constellation, near our home planet, there are a lot of space pirates, and they are looking for any way possible to steal our technology. They are not only space pirates; they are cannibals."

This illustrates how naïve we on Earth are in that we have no idea of what is out there. We may have already offered an invitation to some of these cannibals to visit us. Our ability to breed must make us more than a curiosity! The point he made was that the positive side is always one hundred steps ahead of the negative side in terms of technology. This is why technology is controlled. This tells us that a civilization like ours will never be allowed to possess their technology, until we are in control of our negative side, or until someone comes in and takes control of us.

My visit to the navigational center made it possible for me to understand the technology that enables the mother ship to navigate through space. The technology is based on powerful laser beams that the mother ship emits to search the space skies for a gravitational pull that is in its direction of travel. This book calls a laser beam the "navigational laser locking beam," abbreviated as "NLLB," and the collective system that emits the laser beams the "navigational laser locking system," abbreviated as "NLLS."

The NLLB does not search in the local area for a gravitational pull; it searches in the far away distance, in the direction of its travel. This distance is variable, likely to be in terms of millions upon millions of miles. Distance is a factor because the gravitational pull enables the mother ship to fly at extreme speeds. The power of the force of gravity alone will pull the mother ship at virtually whatever speed they want. All that is needed is a brake, to limit how fast she is traveling.

With this technology, the mother ship is capable of traveling at speeds well in excess of the speed of light. Once a spaceship exceeds the speed of light, it leaves this dimension and can create a dimensional time tunnel, depending on certain variables. (More on dimensional time tunnels is discussed in Book II, Chapter 5.)

When we think long and hard on the last statement, we will have no choice but to ask ourselves the question: How can you leave this dimension, enter a new dimension (which is what happens when you travel over the speed of light), and still have your NLLS connected to the dimension you just left? For, NLLBs are still locked on to the gravitational pull of planets, and they are still propelling your spacecraft through space – albeit a different dimensional space. How this is possible is hard to grasp, and even harder to explain, yet navigational laser locking beams appear to be able to exist in the two dimensions: the dimension you left, and the dimensional plane you are in.

Massive planets are the source of a gravitational pull. When the mother ship comes within close range of a gravitational source it is locked on to, let us say, several million miles, the NLLS (navigational laser locking system) will automatically unlock an NLLB (navigational laser locking beam). All the NLLBs that are locked on will be unlocked in this way. There are always new NLLBs locking on to replace the unlocking NLLBs, and these are locked on to gravitational sources much further away. This is how the mother ship travels through space.

I saw a screen showing the navigational laser locking system. When the mother ship is in flight, colors of lights constantly change from red to amber to green, and vice versa, with each light representing an NLLB. All you see are lights flashing on and off. On or off represents an NLLB locked on or searching respectively. Clearly, not one NLLB operates at any one time; many operate. While some are locked on to gravitational sources, others are in a search mode. The change in color of a light indicates that an NLLB has switched from a locked, a standby, or a searching mode. Depending on the speed at which the mother ship is traveling, you can see which NLLBs are locked on, which ones have just unlocked, and which ones are searching.

A red light represents an NLLB that is presently locked on to a gravitational source. A green light represents an NLLB that is in a search mode. An amber light represents an NLLB that is in a standby mode. This means that an NLLB is on standby to be either unlocked or locked. When the mother ship nears the source of the gravitational pull that it is locked on to, then the light that represents the current locked on NLLB turns amber. Now it is on standby to be unlocked. At the same time, a green light representing a searching NLLB that has found a new gravitational source will also turn amber. Then, the amber light representing the locked on NLLB will turn green, which means that it has gone into a search mode, while the other amber light representing the searching NLLB will turn red, which means that it has locked on to a gravitational source.

Understandably, the first thing someone in the know is going to say is that gravity weakens with distance, especially when you are speaking of the distances referred to here. This is why the mother ship emits long-range navigational laser locking beams, which travel to a gravitational source. For any of this to make sense, we have to look at gravity as a force (forgetting Einstein's ideas and going back to Newton's).

If there are no strong gravitational sources to lock on to, then the spaceship cannot just rely on gravitational pull alone to propel it. This is why the spaceship has a booster system, which is explained below. Usually, if there is a weak gravitational source, then the laser beam will not and cannot lock on to it; it will just pass it by in search of a stronger gravitational source.

There must be at least one question on our minds at this point: What is a navigational laser locking beam? This is not an easy question for a layman to answer. The physics behind the NLLB and the gravitational pull of a planet it is locked on to involves the force of attraction. If you take the force of gravity as being a positive force, and if you produce a negative force (which is what the navigational locking laser beam is), then you form the basis of how the two forces lock on to each other. The aim is to create a force that is counter to the gravitational force emitted by the planet. This counter force forms the basis of an attraction between forces. The laser beam emitted from the mother ship is able to travel vast distances in the universe, and go right up to the source of a gravitational pull. These sources have to be massive planets to have a strong gravitational pull. When a gravitational pull encounters an NLLB, the two opposite forces lock on to each other by a force of attraction. The force in an NLLB automatically adjusts to suit the strength of the gravitational pull it has locked on to, so that the gravitational force of the planet pulls the spaceship toward it. As we know, there is not one NLLB locked on to a gravitational source at any one time; there are thousands, and they can be locked on to different planets. The amount of NLLBs locked on to gravitational sources is variable, and depends on the circumstances.

What is the power behind the force in the NLLB? Nuclear fusion. The mother ship produces nuclear fusion, which is the basis of the power to send the laser beams through space in search of a gravitational pull. This civilization has mastered both hot and cold fusion.

There is a second process involved in this cycle. The NLLB that has locked on to a gravitational pull of a planet is able to extract power from the force of that gravitational pull. This is then collected. It can be stored in the control center of the mother ship, or it can be converted into energy. In this way, the power extracted from the force of a gravitational pull of a planet can be recycled to form the basis of the power that is used to emit an NLLB. As stated elsewhere, in the lower section of the mother ship, in the energy storage center, are reserves of energy to last the mother ship a long time. These reserves are utilized in circumstances that the mother ship cannot find a satisfactory gravitational pull, in which case she relies solely on nuclear fusion to propel her through space.

One last point on the NLLB is that it does not allow anything to come between the mother ship and the gravitational source it has locked on to. This means that the mother ship can never crash into an object in the way. If there are any objects in the way, an NLLB will detect them. Anything expendable may be turned into dust or subatomic particles. Anything not expendable will be avoided. The mother ship will do this by switching to NLLBs that bypass the object.

20
Flying Saucers /
The Physics of Space Travel II /
Security and Combat

Earlier we learned that there are around two thousand flying saucers attached to the base (the control center) of the mother ship. Among these is the fighter flying saucer, which this book has abbreviated as "FFS," which is used for the defense of the mother ship, for missions of investigation, and for reconnaissance duties. Both the FFS and the outdoor flying saucer, which this book has abbreviated as "OFS," have some similarities in shape to the indoor flying saucer (IFS). (On Atlantis, all of these are referred to as a "flying saucer." For the sake of comparison, this book has created separate titles.)

As an inhabitant of Atlantis, you may go on a trip with an OFS. How this works is that you have to hire one, and pay for its hire. You may hire a flying saucer to visit your relatives, who may live in another constellation. You may go on a day outing. You may even go on a holiday. Now we have established the basis of why UFOs are often seen entering into the sea and emerging from it.

Indoor flying saucers can carry up to six people. They are spacious, and are in the shape of a semicircle, although the base is not flat but curved. There are three telescopic legs on the bottom of the saucer. The base of the flying saucer opens, from which you enter or exit it. A strong white beam of light is emitted from it, which penetrates to the ground. As soon as you step into this light, you are automatically transported into the spaceship or out of it.

The dome-shaped top has a solid portion at the apex. Below the solid portion is a window that extends 360-degrees around the spaceship. The final section below the window is a solid material. The OFS and FFS do not have windows. Their exterior is made of the same hard material from which the mother ship is made.

Inside the flying saucer is a map, which is a sophisticated version of our car navigator. Whereas we use a navigator to guide us to a destination, they use a navigator to drive them to a destination. No one can personally fly a flying saucer. A flying saucer is controlled by a computer, which means that you cannot have control of your own flying saucer, apart from telling it where to go. We can look at it as being a computerized chauffeur.

In the case of an outdoor flying saucer, if it performs a 360-degree maneuver, then the pilot will not turn with it, in the same way that our fighter jet pilots do in their aircrafts. An occupant will never feel upside down. He will always feel upright, no matter what the maneuver of the craft, in the same way that no matter where you are on this planet, you always feel upright. We must remember that in space there is no right side up. Gravity creates this illusion. In terms of the IFS, it will never rotate 360-degrees. It has no need to.

Indoor flying saucers have different travel capabilities from the other two types, in that they are not designed for interstellar travel. The technology used by the IFS is different from that used by the FFS and OFS. For a start, the OFS and FFS can travel at speeds that exceed the speed of light. As the IFS is in a controlled environment, it uses antigravity technology, just as the shopping trolley does. The OFS and FFS use the same technology to fly in space as the mother ship uses. Gravity is created by technology in the apex of the OFS and FFS, which has a spinning action.

While in space, the OFS and FFS use the navigational laser locking system, or nuclear fusion, or both together. In an environment of gravity, such as where there is an atmosphere, these technologies are generally not used, although there are circumstances that nuclear fusion may be used. They use a different technology when in an atmosphere: atmospheric pressure.

In the last chapter, we were introduced to the navigational laser locking beam (NLLB), which creates an opposite force to the gravitational force it locks on to. The method of propulsion is a force of attraction. In the atmosphere of a planet, the flying saucer will play on the gravitational force of the planet in a different way. It will create not a force of attraction but a force of repulsion. The same technology to create the force used in the NLLB, which creates a force of attraction, is used in the atmosphere of a planet, only it creates a force that behaves toward the gravitational force of a planet in the same way that two like sides of a magnet act upon each other. The spaceship will create a force that matches the gravitational force of the planet. This will not attract the flying saucer to the planet but repel it. This repelling force is called an antigravity force.

Antigravity technology does not make the spaceship fly in the atmosphere. It can make it hover at whatever height you want it to hover. This means that the spaceship can go up and down, but it cannot move in any other direction, such as sideways, using antigravity technology. Another method of propulsion needs to be used in conjunction with antigravity technology to make the spaceship move in directions other than up or down. Propulsion thrusters are the answer. This is the basis of the technology of the indoor flying saucers and shopping trolleys on the mother ship. Antigravity technology has many applications inside the mother ship; it is the basis of how anything hovers.

To travel in the atmosphere, a different propulsion method is preferred. This involves atmospheric pressure. How atmospheric pressure as a propulsion method works is that the entire surface of a flying saucer has a vacuum belt around it. A vacuum can be applied to certain points of it. When you create a vacuum at a particular point in the vacuum belt of the spaceship, atmospheric pressure will create a force on the opposite end that acts like a motor to the spaceship. The idea is that by creating a vacuum at a specific point on the spaceship, you are also creating pressure, and this pressure is what propels the spaceship. This is an easy enough concept for us to understand.

By this technique alone, atmospheric pressure will propel the flying saucer through the atmosphere at great speeds. In other words, you need to create a vacuum at a certain point in the spaceship's vacuum belt, and this is in the direction that you want your spaceship to travel. Once pressure is created, atmospheric pressure will propel the spaceship. If you want the spaceship to go up, you produce a vacuum on the top of the spaceship. If you want the spaceship to go down, then you produce a vacuum on the bottom of the spaceship. To stand stationary in the air, you only have to produce an equal vacuum on the top and the bottom of the spaceship. Also, the greater the vacuum, the greater the pressure that you create, and the faster you fly.

What determines the direction in which a flying saucer travels is the position of the vacuum. A flying saucer has the ability to not just stand stationary, not just make sharp right-angle maneuvers, but have equal maneuverable capabilities of a fly or other such airborne insect. Now we can understand how flying saucers are able to make sharp right-angle maneuvers. All you need to do is adjust the location of the vacuum around the spaceship. Now we can understand how airborne insects fly as they do.

When it comes to right-angle maneuvers of a flying saucer, you never feel them. You can never become seasick because there is an equalization of forces on a spaceship. Adjusted are not just gravitational forces but other forces that are at play. For instance, you must have the right atmospheric pressure. What we can see is that conditions necessary for human life, such as gravity, oxygen, and air pressure are normal in a spaceship. Gravity is adjusted to a certain weight so that you don't float, or, if you toss a ball in the air, it is not going to take days for it to fall down, or perhaps not even fall down at all. What this means is that a spacecraft has been pressurized to a point that everything has an equalization of forces.

If you consider the case of our lifts, the internal forces of a lift are not balanced. When a lift is in motion, you feel the effects of that motion, which is the feeling you get in the pit of your stomach. On a spaceship, you have the same motions that you have on a lift, even more, but there is a considerable difference: a spaceship has a balance of forces in it, so that you can never feel its motion. Let us consider the case of a bowl of water. If you move the bowl, the water will move with it. This is what happens to you in a lift, because it is not uniquely pressurized for its purpose – which is movement. To eliminate the effects of the motion of the lift you need to seal the lift, pressurize it, and maintain a balance of forces in it. Now that you are in a pressurized lift, you will not know when you are in motion or stationary. If you were sitting at a desk in the lift doing office work, and you were not paying attention to the activity of the lift, and if it were going up and down all day, very fast, you would not know it. If the lift somersaulted and motioned upside down, again, you would not know it; whatever the case, you would always feel upright. If you had your bowl of water in the lift, it would never spill. Having said this, you may have realized that for the physics of this to work there is a condition: the lift cannot be attached to anything. The lift has to hover, because, once it is attached to something, then its conditions become subject to different forces, which are related to whatever it is attached to, such as the building itself.

・ﾉ

Not surprisingly, the FFS is only manned by robots. The metaphysical parts of a governor often take a key role in disputes, and operate through robots. The metaphysical parts communicate with robots all the time. Technology is costly, and wars are rare and always avoided. Humans are never involved in wars or disputes.

The OFS is generally operated by the computer, in the same way that the IFS is. There are prescribed places you can go with one, and the route taken always follows designated flight paths. It is up to you to tell the computer where it is you wish to travel to, and it will program a flight path for you. This is the automatic setting. There is, however, a manual override, and you can take control of the OFS yourself. However, this circumstance is only possible if the computer permits it. The computer reads your mind and knows your intentions. If you want to use the flying saucer for clandestine purposes, the computer will have read your mind and, knowing your intentions, will not provide you with the manual override option.

There are occasions that you might come under attack and need to take control of the flying saucer. On those occasions, the mother ship will usually know of your state of affairs, because the intent to attack will have been detected. The minds of those planning the attack will have been read long in advance of the attack, and an FFS manned with robots will have already been sent to eliminate the target planning the attack; alternatively, the target will have been eliminated by a laser strike that was emitted from the mother ship itself.

How you maneuver the OFS in manual mode is with a joystick. The following is specifically related to the FFS. The OFS does not have all of these capabilities. It does have some, but its capabilities are not as far-reaching. Your front panel has a screen that identifies any objects around you. Radar is the technology used. The radar guides you, and you can use it to lock on to an object. At the end of your joystick is a button that your thumb controls. On the side is another button that your forefinger controls. These buttons fire weapons. Each control has a specific type of firing capability.

One thing you cannot do is accidentally fire a weapon. The computer knows what has to be fired at, so it acts as a safety switch and prevents unwanted fire. If you are in combat, the computer predetermines every shot that is fired and every move that is made. This ensures the success of a mission.

When you see a target, you lock on to it with the radar and press your forefinger. This button is for targeted firing. The thumb button is for general firing, where an emission is discharged from the circumference of the flying saucer. This is to protect you if you have many flying saucers attacking you from different sides. The energy emitted in this circumstance can travel vast distances, and turn everything in its path into dust. Once this is fired, everything around the flying saucer is destroyed. No bullets are ever fired. They do not use such technology. They use laser technology. There is one other weapon used, and it is a particle weapon. When you fire it, it turns the object of its fire into subatomic particles. It only affects matter, so that if there is a person in the craft, his body will be converted into subatomic particles but his metaphysical component will not.

Both the OFS and the FFS use a force field, which protects the craft from attack. This is no different from the force field the mother ship has. Nothing can penetrate it. Additionally, they have cloaking abilities, which means that they can at will become invisible, and never be detected visually or technologically.

If another of your crafts is in the vicinity, it will not face the same fate as your enemy craft, even when caught in a laser-beam emission from your craft. Every FFS has a recognition system as well as counter technology to ensure that friendly fire does not eliminate it. These technologies ensure that they never lose a flying saucer. Such technology is not available to just anyone – only to those civilizations that have a governing authority of the kind Atlantis has. The negatively-inspired civilizations are not given this technology or capability.

Another of the duties of the FFS is to be on guard against enemies and unwanted visitors. The FFS also protects our planet; without this protection, we would have been invaded long ago, and such an invasion would not have been in our best interests. We should note that peace-loving humans would not visit a primitive, non-peaceful race of humans such as ours.

Thus, the FFS often wages defensive attacks on those intending to come to this planet with undesirable intentions. Let us imagine that an enemy craft is on its way to Earth, even at a high speed, with sinister intentions. The FFS can detect this enemy craft and be aware of those sinister intentions long before it reaches Earth. The FFS will then target the enemy craft with a laser strike and destroy it. The Atlanteans have the ability to know what the intentions are of any spaceship coming to Earth.

The FFS is also responsible for the security of the mother ship. No one with undesirable intentions can enter the mother ship. Subsequently, it guards it from external intruders. The mother ship guards against internal intruders. For instance, if someone without permission manages to pass the security system and enters the control center, when caught, he is automatically eliminated – invisible sensors from either the FFS or the mother ship itself will eventually detect him and emit a deadly ray to eliminate him. This applies to spies who don't have physical bodies. Spies from different planets often try to infiltrate spaceships by metaphysical means, which is known as astral travel. In some instances they are not eliminated but captured.

In the universe, there are always humans who prescribe to the negative philosophy of life. These humans only want to destroy. Everyone in the universe has enemies. That there are those who want to steal technology from those who have technology is a part of life. That there are always those who want to have power over others to dominate or destroy is without question. The basic formula of the negative force that exists in the universe consists in the desire to destroy, dominate, and conquer. To those who possess it, technology is highly guarded and protected against falling into the control of those who prescribe to the negative side. Consequently, on the mother ship there is a screening process to determine who is allowed and who is not allowed to go beyond the living center into the control center. This precaution means that there is no possibility that anyone can infiltrate the area. Most superior humans already know their enemies, which means that they know how to identify them.

One type of intruder could be a person who has escaped from a prison somewhere in the universe. This type of person is usually eager to kill you. Escaped prisoners are clever; they have been known to shut down computers and "blind" people of their escape. We must remember that, in the universe, every positive has its negative. No matter how evolved a civilization, it always has a negative side to counter. An evolved civilization always has to be cautious. There is no true utopian society in the universe and there never will be one, for because the universe once had a utopian society and it failed. Thereafter, the universe was purposely given a negative force within it. This tells you that there is no such thing as a paradise out there in the universe, let alone here on Earth. There is never complacency in any civilization, no matter what its degree of intelligence or development. Every living thing in the universe, from the lowest species to the highest species, is always on guard against its predator.

21
A Trip in an Indoor Flying Saucer / Traffic Lanes

I think the greatest thrill I experienced while I was on the mother ship was going to the beach for the first time. Although, there are so many memorable moments of which I could speak. My desire to go there was great, and I was not disappointed.

Guyd was with me wherever I went, and the trip to the beach was no exception. We traveled by flying saucer – a four-man one. The model we used has two seats in the front and two in the back. The back seats are different from the front seats in that they can turn around. By contrast, the front seats are rigid and only face in the direction of travel. Behind the two back seats is the equivalent of a boot.

The ride was so smooth – it is impossible to feel the motion of a craft when it uses antigravity technology. If you were to have a glass of water positioned on a seat, then it would not even shake – this is applicable even if the flying saucer were to turn upside down (which the indoor flying saucer will never do) or make a sudden right-angle maneuver.

Just as we have traffic flow in different directions, so have they, only the majority of traffic flow is in the air. There are no traffic lights or stop signs necessary for airborne lanes. Traffic lanes are always at varying heights so that opposing traffic lanes never intersect. On ground level, vehicles always hover just above the surface of the road, no matter in which direction they travel. In these instances, there are stoplights to allow for pedestrians and intersecting lanes – and there are many cross intersections, so you are not going to drive in a lane over a road with your flying saucer unless it is necessary. The opposite is the case for shopping trolleys and delivery vehicles. Shopping trolleys cannot use the major airborne lanes that flying saucers use. Naturally, there are circumstances that shopping trolleys need to use airborne lanes, such as when they travel to different floors. There are always separate lanes for shopping trolleys to use in these instances. In all other instances, a shopping trolley must use the ground-level lanes.

How the crossings and intersections work is described in Book I, Chapter 25.

There are prescribed height levels for flying saucers and there are designated drive lanes. The computer in the flying saucer drives it for you. No human is able to take control of a flying saucer. There is no manual control option available to the "driver." This is applicable to the shopping trolley.

Guyd said, "This means that there are no accidents. I cannot recall there ever having been a collision. The scenario of either of these vehicles having an accident does not exist.

"There is another factor that makes it impossible for vehicles to have an accident: all flying saucers and shopping trolleys repel one another. This is because they have a magnetic field in them, which is based on two equal negative energies that repel." I understood this. All you have to do is picture like forces of a magnet.

The flying saucer, like the shopping trolley, has telepathic capabilities – in other words, it reads your thoughts. While it may read your thoughts, it does not act on your thoughts unless you command it to, and there is a way for you to give a command to it. You can tell your flying saucer how high you want it to fly, and you can tell it to go faster or slower. For instance, if you are in the vehicle and decide you want to go to the beach, all you have to do is think about the beach and command it to take you there. Along the way you can tell it to alter the speed at which it travels, and it will follow your instructions. Naturally, there are speed limits. While you can request the flying saucer to exceed a speed limit, it will not act on your command. Hence, you can only command it to perform logical maneuvers that fall within a range of preset variables.

If there is too much traffic in the lane you are in, and if you feel that you are traveling too slow, you can tell the computer to change levels, since there are several levels of traffic that move in the same direction. There is not one set lane. The flying saucer travels in an orderly way, with other vessels in front of and behind it. Therefore, you tell the computer to go up a level, but you only do this if you can see that the lane is moving at a faster pace. Your flying saucer will then find a space in that lane. You can let it take you to your destination on its own, or you can take control of how fast you travel, which lane you travel in, and so on. However, you cannot operate the indoor flying saucer on your own.

Air roads work on levels. You can see other vehicles going in the opposite direction, but they are never close to you, and they are usually at a different height. There is no stopping on roads, either. What I saw were congested lanes of flying saucers.

As we traveled, I was paying particular attention to the other people traveling, to the shops below, and to the scenery. What struck me most was that everything was always clean and beautiful to look at.

We arrived at the beach. In the near distance was a parking lot. Flying saucers were parked above one another, in empty space, with no floors or walls between them. Our flying saucer stopped at the beach to let us out. Guyd said, "It is going to go and park itself in the parking lot until you are ready to leave. You won't even know where it has parked. When you are ready to leave, all you have to do is request it and it will come and collect you."

They carry a remote control device with which they can summon their flying saucer. When you press one of the buttons, lights on the flying saucer flash. Guyd demonstrated this to me. It works in just the same way that our car remote controls work. This means that if one method of calling your flying saucer doesn't work, you always have a back-up method. Things can go wrong with a flying saucer, and it may not be able to tune in on your telepathic frequency. If this happens, then your back-up device will come in handy.

There is always a possibility that technology can malfunction. In the circumstance that you cannot operate your flying saucer, you have to call for a technician. This is the same technician you call upon when you have a medical condition. You don't have to pay for his services, in case you are wondering. He is going to send the malfunctioning flying saucer to his workshop. He won't tow it, by the way: you won't see a flying saucer towing another one in their society! He simply presses a button on his gadget, and the flying saucer disappears and instantly reappears in his workshop. This device uses subatomic particle conversion technology; matter is converted into subatomic particles, and then the information on its blueprint is transmitted as signals directly to a converter in the technician's workshop.

You may wonder how it is that the person whose flying saucer broke down then goes about his business. In answer to this, the technician gives him a replacement flying saucer. When the old one disappears a new one appears, and he keeps this one for good. Thus, no one is ever left stranded if his flying saucer breaks down.

In the parking lot that day, there were ten levels of flying saucers. There was approximately seven feet of space between flying saucers. There must have been thousands of flying saucers there, which led me to believe that the beach is a popular attraction.

22

The Beach / Sexual Attraction

The beach itself is approximately five kilometers long; it looks real in every sense, and even has waves that wash ashore. From the shoreline, the sea fades into a distant horizon. But this is only an illusion, for, the horizon ends about two and a half kilometers out from the shore.

While there are no boats in those waters, there certainly are surfers. The highest wave I saw was around ten meters high.

One thing that immediately caught my attention was that I could not see a lifeguard. "We have no need of lifeguards." It was funny when Guyd added, "The lifeguard job belongs to the dolphins." That certainly got me wondering. "If you pretend you are drowning, one is going to come after you, get you on your back, and in no time at all toss you onto the shore."

To myself, I thought, "Friggin fish! Must be smart."

Thoughts however, are never private. Guyd smiled and said, "You are a funny man."

"Well, I've been a clown in the theater, and they used to call me a clown, but never a funny man."

He laughed and put his hand on my shoulder as we walked to the beach.

The first thing I did was take my shoes and socks off, which I left on the sand. I had to roll up my trousers. I was casually dressed in trousers and shirt. I left Guyd behind and went up to the water to taste it. Yes, it tasted salty! Frankly, it was the first real water I had seen, apart from drinking water. I noticed a multitude of marine life within it.

I walked back to Guyd and took a seat beside him on a bench. The bench was not hard like timber; it was as soft as a cushioned lounge, yet it looked like timber. I said, "This is ideal for my tushy – very soft. I hate sitting on hard surfaces."

Guyd laughed and said, "Your tushy must be very sensitive."

"I know. I only hope you don't have a red-back spider hiding underneath."

"Jack, we don't have red, blue, or white spiders on the ship! The blue fog kills anything like that." He grinned and shook his head, and then added, "Besides its recreation facilities, the sea offers a rich supply of food, and a great deal is harvested from it. As you proved to yourself, the water you see on the beach is real. When the mother ship is not docked somewhere, as it is now, but is roaming through outer space, we don't necessarily keep seawater in this form. We can convert it into tablets that are lightweight, and are only around one and a half inches in diameter. This is done when we have to conserve weight while traveling."

I wondered where they keep dolphins when seawater is converted into tablets, to which Guyd responded, "The mother ship has four strategic areas that contain water. These areas are designed to balance the mother ship. Dolphins are simply put in one of these four locations, in just the right numbers; then they are manually fed."

I could see dolphins being ridden on by children, and water skiers towed not by boats but by dolphins. When I wondered whether they had whales, Guyd responded, "Dolphins are the largest of the marine life we have. One reason we don't have whales is that it is possible for one to kill someone with its tail; the other reason is its weight."

I saw scores of beach umbrellas. The temperature was around thirty degrees Celsius. People were sunbathing. Some were playing with kids. Some were building castles on the shore. There was nothing different about what I saw there and what I would see on one of our beaches, apart from the dolphin activity.

I saw some women in bikinis, but most were wearing one-piece swimming costumes. One thing that stood out for me was that I saw no overweight people. The diets and metabolism of these humans do not provide them with the opportunity to become overweight. Every single person I saw looked no different from us. Some people had short hair; others had long hair. Men were an average of six foot one in height, while women were an average of five foot eight in height.

"As we develop the rest of our brain, our looks will slowly change, as will yours one day. Presently we use three-quarters of our brain; we hope that the next generation is able to have the full use of their brains. When this happens, hair will no longer grow on their heads. Hair is a parasite and does not serve a useful purpose," Guyd said.

I noticed a nudist section further down the beach. This was completely unexpected. Guyd read my thoughts and said, "I've got no problem with it! I've got used to it."

"Of course you don't! You're a robot. You don't have the right tool for this business."

He laughed and said, "You're a funny man." That was a regular comment he made to me. Then he added, "No man or woman is sexually stimulated by the sight of a nude body."

"Why do they do it then?"

"They just want equal rays of the 'sun' to shine all over them. In a swimsuit, some of these humans feel as though they have a raincoat on. Whatever their reason, it is not sexual; these humans don't have sexual perversions as your humans on Terra have.

"Sexual attraction between a man and a woman of this intellectual caliber can only happen when two intellects 'lock' on to each other. There is no sexual attraction as you know it; there is a mental, intellectual symmetry between a couple. When this happens, they will talk to each other telepathically, and then the couple will hold hands, so to speak, from then on. That is all they need to do to know that they are compatible. This is how they form an attraction."

I understood that an attraction between a man and a woman in their society is not based on a man seeing a woman's figure and vice versa; it is a mentally based attraction. Guyd interrupted my thoughts and added, "The younger generation of women does dress to attract the opposite sex. These women are looking to attract a husband and have children. Once they marry they change, and then they behave as women should. Besides, if a woman happened to be dressed provocatively, no one would be enticed by her, as sex does not sell in this society.

"You are right in that humans from planet Terra often have problems, because many fall in love with what they see, based on looks and sex, not on intellect."

I wondered at what age they actually do marry, considering the term of their lives, and Guyd was quick to respond with an answer. "They marry anywhere between the ages of 21 and 120. And they don't cheat on each other. In an advanced society, you cannot hide anything because you can read everyone's mind and know everyone's secrets and habits. If you cheat then you will have degraded your intellect. All that you are is your intellect, and what you value most is your intelligence, and this intelligence is what emanates from you when you talk and breathe. The women who are pursuing a husband may dress provocatively to a degree, but they don't have any partners or relationships before they marry."

I understood that intelligent people go back to traditional values. Guyd agreed and said, "The first partner they find will be the one they marry. They know in the instant that their eyes magnetically lock that they are capable of forming a lifetime partnership, and there is no such thing as divorce." I found this astounding, in view of the length of their lives, which we know can extend well beyond 1500 years.

One thing I noticed was that women did not wear make-up – in an intelligent society women don't. They rely not on makeup to look attractive, but on an intellectual aura that is generated from their intellect, and that aura can only come from having a developed intellect. The women are always natural in appearance, and their intellectual aura makes them attractive to the opposite sex. Furthermore, it is only natural that their skin would be in perfect condition in a world that is disease free. They have the smoothest, most tender, pinkish-cream skin I have ever seen. I never saw a person there who was not physically beautiful to look at.

The other thing that I noticed was that neither women nor men wore jewelry. There were no rings, no earrings, and no watches! Even when couples marry, they do not exchange rings. I was told that wearing jewelry is seen as a primitive custom. To these people, wearing jewelry is generally indicative of insecurity. These people have nothing to feel insecure about. For a start, there is nothing in their society to destroy them. There are no crooked governments. There is no job insecurity. There is no mortgage stress and there are no banks ready to evict. There are no home evictions. There is no such thing as battling for money to feed your family. There are no neglected elderly and vulnerable. People don't take advantage of others. There are no disadvantaged people. There are no crooks, thieves, liars, or cheats.

Just as there are no jewelry shops, there are no tattoo parlors. The psychology behind wearing jewelry applies to having tattoos. The image the people of Atlantis present is an intelligent one, based on the aura they project, which is directly proportional to their intelligence. They do not need tattoos or jewelry to create an image.

I noticed that trench coats were a common outfit worn by men, and this was worn with hats from the forties and fifties era. In general, the fashion they wore while I was there was from a mix of eras. I saw people who were wearing clothing from the forties, alongside people who were wearing clothes that you will see in our present era. I didn't see anyone wearing sunglasses. The so-called sunlight emitted in the mother ship does not possess the damaging qualities that the rays of the sun possess, from which you require protection. It is a different story when they visit us on the surface of our planet. In this instance, they don't wear ordinary sunglasses. The top lens offers protection from the sun in the same way that our sunglasses do. The bottom half acts as an x-ray. If you see someone tilt his head up around two inches as he talks to you, he might just be an Atlantean scanning your body! There is only one reason for this – to see if you have weapons on you that could harm him. They mainly wear these glasses when they visit us or other planets.

23

Dolphin Surfing

Many shops line the beach. What impressed me was that the shops on the streets are as up-market as the shops in the shopping centers are. I could not see any inconsistency in their appearances as there is in our society, with street shops often rundown and neglected. Every shop that I saw was maintained in such a way that it presented itself in the highest possible standard. I did not even see litter, and I did not see one dirty footpath or street.

I saw lots of paths and walkways around the beach, which had many kids riding bikes on them. The bikes these kids were riding are a little different from the bikes our kids ride. The tires are around three inches thick, and in the middle is a gap. Their bikes have unique qualities, one of which is to maintain equilibrium. What this suggests is that a bike does not allow a child to fall off it. This is why helmets are not worn. I didn't see any skateboards. Parents there are extremely protective of their kids. Children are so intelligent; their brains are like computers. A kid in their society will never fall over and sustain an injury. No kid in their society ever has. Even the shoes children wear are designed to maintain balance, which means that they do not enable a child to fall over. If the circumstance arises that a kid is about to fall, his shoes will rebalance him so that he does not.

Two other pastimes I saw kids indulge in at the beach were to ride the surf on the back of a dolphin and to be towed on a surfboard by a dolphin. Guyd said,

"These activities are completely safe because a dolphin will never let his rider drown." Children were not the only ones riding dolphins: adults were participating in this activity. "In the case of an adult, there is a condition – a dolphin expects to receive a fish as payment," Guyd said. Now I understood why I saw so many local fishmongers. What I found fascinating to see was an operator running a taxi service at the beach. I suppose there is nothing extraordinary about this, until you discover that he was using dolphins to ferry people from one place to another.

Incontrovertibly, their dolphins are exceptionally intelligent, and know just how to entertain you. On the scale of intelligence, the dolphin ranks next in line to the human. Because our telepathic capabilities are not yet developed, we cannot fully appreciate the intelligence of the dolphin species. Dogs cannot match the dolphin in intelligence. Once you have telepathic capabilities, you can communicate with a dolphin just as you can with a human being.

There were people everywhere, many enjoying an ice cream. There were sea gulls everywhere – according to Guyd, they are always there because there is an abundance of fish to attract them to those waters. Numerous species besides the dolphin feed on the fish. What I found to be the funniest thing ever was when I went to give a dolphin a fish, and he said to me, "No thanks, I'm full because I've already been hunting and have eaten so much fish for dinner. So eat it yourself!" I was laughing to the point of hysteria. I discovered that dolphins are full of jokes and have a quirky sense of humor. I spent some time merely observing their behavior: they were showing off and performing aerobatics; at one moment they rode the surf to the shore and came right up to the beach. I felt that they were taking a bow for their performance when they stood upright in balance and made their typical vocalizations. They almost appeared to be laughing and clapping, and then in one swift move they disappeared back out to sea. Although it looked as though it were a trained performance, Guyd said that it was of their own volition that dolphins perform.

I could not resist the temptation to take a swim. Guyd said, "Leave your clothes here and I'll look after them."

With only my shorts on, I walked toward the water. I felt a little self-conscious because everyone in the vicinity was staring at me. Everyone could tell I was different. Even so, I was glad that they were intelligent enough not to bother me.

Once in the water, I could feel that scores of sardines were around me. It was thrilling to have a school of them under me, supporting me. I outstretched my legs and arms and bodysurfed on them.

The water was so clear that I could see to the bottom. I was in water up to my waist, intending to find a dolphin to ride. I had the pleasure of encountering a most unusual one: one that outsmarted even me! I called out to one telepathically, "I need a ride." That was when a large fin and a dark shadow that reflected a touch of white on its sides came toward me. My instincts kicked in and I panicked in that moment. Thinking it was a shark, I raced back to the shore. It was funny because no one ever thinks along such lines. In this respect, I was unique to the dolphin, which was its incentive to come after me. I couldn't make it back to the shore; he was too quick for me. He plunged himself between my legs and tossed me into deeper water. I could instantly feel the change in its temperature; however, it was still pleasant. I was no longer in a state of panic because I knew I was dealing with a dolphin and not a shark. For a start, I remembered what Guyd had said about there being no dangerous creatures in this place. Suddenly something appeared in front of my nose and nodded its head. It was the dolphin; he was only about twenty inches away from me, making a clicking sound.

Waves kept crashing over me as I was treading water. I kept going underwater with every wave. I have to admit that I was a little scared. Then, after a new wave hit me, I telepathically said to the dolphin, "I want to ride you. Get me out of here!" This time he came under my legs and took me for a ride – there was only one problem: I was riding backward, just as cowboys did on their horses in the Wild West. He kept weaving in and out of the water, which meant I was going under often. My head was even going under. This was not pleasant because I had nothing to hold on to and he was slippery. It was no surprise, then, when I fell off him.

He came back to me and went under my legs the right way; he paused long enough for me to grab a hold of his fins. He told me exactly what part of his body to hold on to. He was then surfing on top of the water with his head down. At one point he lifted his head out of the water and said to me, "Where do you want me to take you now?"

"Closer to the beach; not the deep end." As I neared the beach, I noticed that I had captured the attention of around one hundred people. It appeared that I was the object of their laughter. Once I arrived closer to shore, I tuned in on what they were saying. All were wishing for a dolphin to ride such as the one I had – his unusual sense of humor impressed them. It certainly did not impress me.

The dolphin responded to their comments by turning back and saying, "You'll never find anyone like me. It will cost you a lot of money!" He made an audible noise with his mouth and then turned around and took me for a ride all over the place. I remember asking him, "What money? What do you know about money? You can't go to the shops and buy fish!"

"No, they have to bring me a ton of fish – live ones, not dead ones, so that I can gather my friends – I have hundreds of them – and have a feast. Besides, Jesus Christ could make a ton of fish from one fish in a matter of seconds. It didn't cost him anything. They can do the same by just using the food converter. And make sure that the fish don't have bones. There are fish here that are not only smart, but also a menace. They have too many unnecessary bones that can clog the food in you as you are processing it, which can kill you."

"But a ton of fish? You'll be like a balloon!"

"No, don't worry. All my friends will clean the place up within ten minutes. They are all hungry. They don't know how to fish. I know them. They are all youngsters mainly, just beginning to be taught by their parents. You have a shoal of fish coming up. Look at them; they are jumping up out of the water to get oxygen. It looks as if there is too much salt and not enough oxygen in the water. I'm quite sure the owners of this sea will rectify this very soon."

"I'd like to go back on dry land now."

"The beach is not dry land; it is built on a dry place!"

"OK, never mind, you just hurry up. I've got a friend waiting for me on the beach."

"OK."

I regretted hurrying him, because he swam to the beach so fast that he scared me to death, and then when he arrived in shallow water he made a sudden stop that caused me to propel over his head and onto the shoreline. I was not impressed. I even had to go back into the water to wash the sand off me. I noticed that the dolphin was amused. He was in an upright stance, laughing and clapping. Then he did an aerobatic maneuver, before clapping and laughing at me. Then he repeated the aerobatic maneuver. This was his way of expressing pleasure at what he had done to me.

I took a handful of sand and threw it at him when he balanced in an upright posture. Needless to say, he dodged it by immersing himself in the water.

Once I stepped out of the water, he called out to me telepathically and made an audible sound. He said, "Hey."

I turned around and there he was. He said, "Don't forget that ton of live fish – preferably sardines."

"Yes. I'll have a lot of questions from all these people. I'll tell them to fish out one ton of sardines and drop them out there for you."

"Fifty meters from here; I'm always there."

"I'll definitely do that. I will tell everybody, because you scared me."

"I will perform for them all if they do that. But they won't come back for another ride, because they're all cowards, like you!"

He was right in saying that because I would never go back for another ride on him. I couldn't breathe underwater for as long as he could; thus, I panicked. In those moments that I couldn't breathe and panicked, I could hear his laughter in my mind.

He added, smugly, "Yes, I bet you there won't be many, because they are cowards!"

At that moment, people ran up to me and asked me what was happening. I heard all their voices in my mind at once. Someone said, "I've never seen anything like it." And, "How did you get a dolphin like this? Tell us! Tell us!"

I said, "You collect some money – however much it costs to purchase one ton of live sardines. Then when you have the sardines, drop them down there in the ocean for him, fifty meters from here. If you can do that, he said he's going to give you all a ride, just like the one he gave me. That's his business. He's blackmailing you; for fish, he'll entertain you. That's what he told me to tell you." Without hesitation, many rushed off to the shops to do just that. They all wanted to experience a ride on that dolphin.

It appeared that this dolphin stood out from the other dolphins. I did not experience any other so I had no way of making the comparison, but it seemed as if this one was too smart. He was a joker, and knew how to converse well with people. I was expecting to have a ride on a dolphin, but not on one that reads my mind and deliberately plays tricks on me, according to my fears.

What astounded me most was his degree of intelligence and knowledge. To know about money is one thing, but to know about our historical facts is another thing altogether. If I did not know better, I would have bet my life that he had had an education.

I met Guyd on the beach. He just appeared next to me. Whenever I thought about him, he appeared. He told me that he was nearby, sunbathing. I did not recognize him, and never would have. He was without his regular hat and suit, but was in his swimming costume. It was the first time that I noticed he was bald. Even though he was a robot, he acted just like a human. It is hard to imagine that a robot without a soul could have not only the feelings and understanding of, but also looks that are indistinguishable from, a human.

The first thing I asked Guyd was, "Who teaches these dolphins historical facts?"

"A lot of people come to the beach. Some lie on the beach and read books for hours. The books they read are from all over the universe, including from your world. As the human reads, dolphins pick up their telepathic signals and absorb that knowledge. This is how dolphins learn from humans. They know history; they know everything! In reading your mind, a dolphin is even capable of extracting knowledge from you, and, sure enough, he gains much knowledge by conversing with you. He also spies on people, and listens in on their conversations while they are at the beach. He picks up much knowledge this way."

While conversing with Guyd, I learned that several species of fish are almost as smart as dolphins – particularly catfish. These species are able to communicate telepathically. Any species that is able to communicate with you telepathically is a protected species, and it is not to be caught or eaten. Any species that can communicate telepathically with a human is in some respects considered synonymous with the human species, even though it has a different life cycle.

In the sea world, bigger-sized fish generally have a higher grade of intelligence than smaller-sized fish, which has logic to it. For instance, an intelligent species of fish, such as the dolphin, will never eat a species of fish that is intelligent and can communicate, in the same way that a normal human will never eat another human. An intelligent species of fish is able to distinguish between an intelligent species and a non-intelligent species. This then determines the eating habits of an intelligent species, and the placement of intelligent life in the food chain, which is high up the order. It appears that in the sea world on Atlantis, intelligence is directly proportional to size. It is logical that the smaller species of fish should be stupid so that the smarter species of fish don't have a predator. Of the smaller species, sardines are in constant supply as they multiply rapidly; thus they are the prevalent food source.

·᷄

A memorable moment occurred when we left the beach. At first I thought that the loud noise was from someone calling out to me. Guyd said, "Look back, your friend wants to say goodbye to you." I turned back and saw the dolphin I had befriended earlier. He was bouncing upright in the water and flapping his fins. In his mouth was a sardine. It was an unbelievable sight. Then he spun around, like a whirlpool, and with momentum he swung the sardine that was in his mouth with such accuracy that it hit me on the head. He was about twenty meters away! I heard the dolphin's laughter in my mind. Then he disappeared in the water.

Everyone at the beach was clapping at the dolphin for hitting me with such accuracy. These dolphins have a one hundred percent accuracy rating when they set out to toss something. I picked up the sardine; when he popped up out of the water he was still laughing at me, so I tossed the sardine back at him. My aim was off, which inspired him to try to catch it. Then I could see him with the fish in his mouth again, and I understood that he wanted me to play catch with him. He was going to propel it again, when I said, "Oh, I've got no time to play with a fish!" Then I left, unimpressed. However, we had only walked a short distance when I felt a sharp jolt of pain at the back of my head. I turned around and saw a sardine on the ground. I looked into the distance, only to see the same dolphin, who was laughing hysterically. I took the sardine, which was larger than the last one he threw at me, and ran toward the water with it. The dolphin just looked at me; then when I threw the sardine, he dove underwater. Amazingly, he retrieved the very sardine I threw at him, and wanted to throw it back at me.

I shouted, "You eat the fish, otherwise you won't get any next time. I'm going to throw the next one in the rubbish." I could tell he was offended: he just dived into the water, and when he came up he didn't say a word.

24

The Bird Life

There are many species of trees and plant life on the mother ship. I saw blueberries climbing up trees, as though in quest of a sun. I saw mushrooms budding from a productive soil, of which worms are an industrious aspect. The vegetation has a cycle of life and death, and the span of life in both the plant and the animal world corresponds in many respects to the term of life of the human population. This means that everything lives for a longer spell on the mother ship than it would on Earth itself. In that cycle, some species experience seasonal change: they flower, fruit, and even shed their leaves before a period of dormancy, even though there are idyllic climatic conditions on the mother ship. That there is only one constant temperature on the mother ship suggests that plants are genetically altered, which allows them to adapt to the environment in the way they do. Additionally, while there is a seasonal cycle, it is not a twelve-month seasonal cycle. The cycle is shorter. As there is no rain on the mother ship, the vegetation extracts its water needs from an underground water source; where it pools at the surface, it provides a means for life to thrive, such as the birdlife. Birds are plentiful, and their songs stir in the otherwise still air.

I felt so at peace as I walked through this lush paradise with Guyd. Little did I know what I was about to encounter. Guyd was obviously looking forward to seeing me get myself into trouble – he was so sure I would. He found the entertaining moments I provided him with priceless, because they don't exist in a society where ignorant people and hence stupidity are rare.

The birds have evolved in an inimitable way in their society. In the package of their intellectual evolution are quirky personality developments. Intelligence is one thing, but personality is another. One thing I came to learn, the hard way, is that for all their cleverness, birds, particularly parrots, are temperamental and touchy. This means that you have to be careful of how you speak to birds. It took me an encounter or two to realize that you have to flatter the birds to get along with them. Birds will often perch on your shoulder. Some will insult you by saying (telepathically, of course), "You don't speak very well." If you happen to tell one something that hurts his feelings, he will take off and leave you, but not before expressing his feelings by leaving a dropping on your shoulder. This is his way of thanking you for your comments.

My first incident involved a parrot. One came along and perched on my shoulder. My natural instinct was to think, "What do you want on my shoulder, you stupid buster?"

He reacted by repeating my words, in a squealing sarcastic way. Then he took off and left his "signature" on my shoulder. People around us were watching, listening, and laughing. Some birds have a habit of perching on your head; in this instance, you had better be careful of what you say. Indeed, you have to patronize them by telling them how beautiful they are. After my first encounter with a bird, I was sure to treat the next bird I encountered in a very different way. One landed on my head, so I said, "Did anyone tell you how beautiful you are?"

Perhaps it was the way I had said it. Perhaps I sounded too patronizing. Whatever the case, he did not find my comment agreeable. He responded, "Thank you." He rose in the air and added, "My foot, beautiful!" Then he left his large signature on my head, and it was not pleasant at all. Unsurprisingly, I was entertaining to those who were watching me. I know that they have never encountered someone like me in their midst. Anyhow, I had to make a quick withdrawal to the beach to wash my head and shoulder. I discovered, on another occasion, that some birds nibble on your ear just to annoy you. Birds with this personality trait also have a habit of leaving a signature behind if their egos are bruised in some way. From then on, I made a point of distancing myself from birds.

25

The Streets

There is appeal in everything that you see on the mother ship. No attention to detail has been overlooked. Every evening the so-called suns that illuminate the mother ship are dimmed to create the illusion of evening. An illusion of dawn and twilight is also created to denote the beginning and the end of the day respectively. After twilight, dark on the mother ship is akin to a night with a bright moon. To add to the ambience, trees shimmer with fairy lights. Rows of columns bear three luminous lamps. Without sounding repetitive, the scene is breathtaking, and it leaves Paris at its best as a poor imitation. One cannot envision the atmosphere. When you walk along the streets – whether by night or by day – you experience a moment that you can never forget, and you can never find words to describe what you see or feel.

The quality of humans complements all of this. That it is rare to find people of such caliber here on Earth is an understatement. Anyone you walk past will exchange a greeting with a nod and a smile; many will say hello to you telepathically. What I found most appealing is that everyone there is intelligent; there is no place for stupidity and negativity, as we know them. There is no jealousy or envy. Having said this, because these humans have not attained equilibrium (as described in another book), they occasionally fall into the trap of the negative cycle, but their negativity is nothing compared with our negativity. In their society, there is no need for psychiatrists or psychologists. That said, it is easier to remain on a positive course when many of the negative-inducing circumstances that confront us on Earth simply do not confront them. Appropriately, these wise words tell the story . . . If you give a maggot everything, it will leave you alone and be the perfect maggot; change its environment, and that maggot will not just turn on you, but feed on you!

In view of this, humans still have the odd flare-ups, and there have been the occasional brawls, but not as we know them. Sometimes humans have been known to have a punch-up because of a disagreement. But there are no murders. In a place like that, you know that the people are for the most part calm, intelligent, and happy. They all have a capacity to reason. In my short stay there, I never once witnessed a disagreement or even heard a bad word. Furthermore, you never have to put up with people presenting one face to you, and then behind your back presenting their true face. This type of behavior does not exist in a place like Atlantis. The possibility of this has been negated by telepathy. It is hard to imagine what it means to be around real people: people who do not lie, who do not cheat, and who truly care for and respect others.

You cannot compare these humans with the humans that we have on the surface of our planet. For this reason, the Atlanteans would have more in common with some of their native creatures than with most of us humans. Indeed, some species of wildlife have a higher IQ than many of us have, and offer them an intelligent conversation when many of us could not.

When you walk along their streets, many of which are French in style, you cannot see a trace of graffiti. I was wondering what the Atlanteans would do to graffiti artists if they caught them on the mother ship!

Guyd responded to my thoughts, "It would be a disgrace to our culture to have graffiti artists. If we ever did, we would just wipe the negative desire from their minds, so that they never think of doing anything like it again. This can only happen when someone from another planet comes here – someone who is totally stupid. I know exactly what I would do to him, because I have done such a thing, for doing the very thing we are discussing. First, I would make sure that all the people here look at him as being some sort of invader, and take the precaution of keeping clear of him because they would know that he's done something wrong.

"Once, we had one such visitor, who put his graffiti all over the place. Once I caught him, I just pointed my finger at him and made him fly up to the ceiling

The Streets

– very fast, mind you. He nearly hit the roof, but I stopped him right on time. Then I turned him over so that his head was pointing down. He screamed and yelled. Then, with his head still pointing down, I made him come down like a rocket, very fast. I stopped him just a few feet off the ground. Then I made him somersault so that he landed upright on his feet. You should have seen him run, like a jackrabbit. Not even a rocket could have caught him. And we definitely got rid of him soon enough.

"Everyone who left the shopping center saw the graffiti and couldn't believe it; those who were around while I was having fun with him stopped to watch. They could not stop laughing. Their laughter became hysterical when they saw how he ran off – as though he had a motor on his backside. It was one of the most entertaining moments we have had on the mother ship. Naturally, such things don't happen in our society. Only when an outsider comes here, can something of this nature happen.

"Seeing him run off was the most picturesque scene I have ever seen in my life. It was a pity that I didn't have a video camera with me. I would have taken a video of it and made a lot of money. Everyone would have loved to buy a copy of it, because it is not often that you see such funny things. Things like this emanate from stupid people; because we are not stupid, we don't do or see stupid things.

"As for graffiti, it wastes time and is costly. From time to time, for instance, every 100 years or so, we will catch an outsider, someone from a different civilization, from somewhere in the universe, who comes in here and shows us what he can do, which to us can look very funny or not funny at all. Whatever the case, it looks stupid. We don't entertain people of that nature, so we just send them back to wherever they come from."

Many of the footpaths and walkways are paved with tiles that are non-slippery and made of industrial gold, which is identical to gold. The finish is a polished-looking matt gold, which never scratches. As you walk along the footpaths, often you see people passing by in shopping trolleys, who are going about their business. Their shopping trolleys hover at least a foot off the ground. There are no line markings on the road; they are not necessary, as all vehicles are driven by a computer, which negates the need for visual guides on roads.

At every intersection, on the footpath, is a mosaic, which is the shape of a star. There are no street signs that we are familiar with, for, the mosaic is a form of computerized signage. This means that if you find yourself lost, you only have to go to the nearest intersection, to the mosaic. What is interesting is that the language in which the information in the mosaic is written can adjust to the language you speak. Obviously, it possesses intelligence to be able to read your mind. You can have one hundred people reading the information, and every person is capable of seeing it in a different language.

The streets seem to be straight, and are all set out in a grid-like fashion. Everything is symmetrical and architecturally planned. The intersections are interesting in that there are no pedestrian signs and no traffic lights. In the middle of an intersection is a colored marking, which is often olive; this marking is in the shape of a circle. As we know, vehicles hover just above the ground, so there must be pedestrian crossings. In addition, on ground level, every direction of traffic travels at the same height, which means that one lane of traffic must give way to the other at cross intersections.

By law, pedestrians are only permitted to cross a road where there is a circle marking on it. While vehicles hover at a slow speed, there is the possibility of injury. The crossing that you use bisects a circle marking on the road. In our world, the pedestrian crossing is often represented by yellow stripes marked on the road. Instead of such markings, they have a light that is emitted from the road. The alternating light is green, orange, or red. Obviously, a green light indicates that you can walk across the road, while a red light indicates that you cannot. There are sensors at intersections that detect the presence of someone at an intersection.

The pedestrian light is not visible to vehicles, just as the light that is emitted for vehicles is not visible to pedestrians. Realistically, the lights at a crossing are for the sake of pedestrians. Vehicles do not need any lights to know when to stop or go, which means they do not rely on the lights in any way. The sensor at a pedestrian crossing alerts the main computer of a pedestrian. The computer is linked to every vehicle – whether that is a shopping trolley or a flying saucer. This means that the main computer controls the movements of vehicles at an intersection.

In the dimmed light of night, the gold paving of the road looks magnificent against the backdrop of street lamps, and against the trees that glisten with fairy lights. The fairy lights bear no resemblance to the ones we are familiar with; they are industrial-grade diamonds, and each diamond is illuminated. The gold surface of the road reflects all of these lights. In this place, 1500 years is not long enough to live.

The antique-looking street lamps are either French or English in style. There are generally three lamps on each lamppost, and the bulbs emit an opal-colored light. The lampposts are olive in color, with gold vertical stripes. You can never see any dust in the mother ship, so everything you see looks brand new. This can be attributed to the blue fog.

One thing that I found curious was that the people love to see visitors. Everyone whose path I crossed instantly knew I was one. People didn't have to be introduced to me to know that I was a stranger in their midst; from afar, people often stared at me. Often I drew crowds of curious onlookers. Many wanted to meet me, while others were happy just to see me. Everywhere I went, I was treated as special; often I felt as though I were a visiting monarch. Guyd repeatedly introduced me as being a man from planet Terra. In response, everyone was pleasantly surprised, and even excited at meeting me. I was always asked questions. People were naturally interested in what I had to say about planet Terra. One thing that didn't escape me was that there are some very beautiful women there.

I remember how one person in a crowd asked me about my diet back home. "Do you eat fish?" he asked.

"Yes; we also eat snakes!"

Everyone laughed and mumbled, "Yuk!"

"We love crabs. We cook them up and make a nice dish out of them," I added.

"Well, the same as us, served with a sauce," one said.

Another asked, "Would you like to stay here with us?"

"Oh yes! I would – if I find a beautiful woman who is prepared to live both on the land and in the sea."

After this remark, I noticed that many women were weaving their way through the crowd toward me. They were dressed in different styles of clothes, such as miniskirts, maxi skirts, or pantsuits – some with many zippers. Some of the women were eating ice cream. All were nicely groomed. When I saw their interest, I added,

"I'll probably become a double Muslim." In reality, I felt as though I wanted to run away from them all, as I got a little spooked by them.

Everyone laughed.

All this laughter attracted the attention of others in the area, who stepped up to the crowd. One man asked another man in the crowd, "What did he say?"

He told him what I had said, and, in response, the man who asked the question opened his mouth in astonishment and covered it with his hand. Then he turned his back to laugh in private; others nearby were listening in, and they covered their ears and laughed. More people joined the crowd and asked, "What did he say?" When they heard my comment, they too showed a sense of modesty. All were surprised. I don't think that these people would have ever heard a dirty joke in their lives, and they would not have met a man like me. I had fun answering questions. You might say that I was in my elements with these people, answering questions with stupid answers. The stupider the question, the stupider my answer.

I learned much about the people. For instance, many had emigrated from different constellations to the mother ship.

Guyd and I walked along a main street, with a crowd of people now following me. I had two young women on each side of me, and we walked arm in arm. I remember boasting about myself, telling them that I was an actor.

One of the women said, "We also have actors, and we have regular performances in theaters. We also have cinemas, just as you have."

As we walked, I looked at one and said, "You have a nice nose." I turned to the other and said, "You have a nice upturned nose, and look like a Scandinavian woman I know from back home."

"Many people have told me that I have Scandinavian looks." Then she added, "Where is Scandinavia?"

"When you get out of here, go up north to the coldest place on planet Terra – where you'll freeze to death if you are not well dressed!"

"Do you have people like us on planet Terra?" someone asked.

"We have a lot of stupid people. There are lots of good people, but there are lots of bad people. There is no harmony, and sometimes people declare war and brother kills brother." All just shook their heads, finding it difficult to comprehend.

"What kind of people can do that?" someone else asked.

"We have different people, and we have people with varying degrees of intelligence. Intelligence is the factor that dictates such behavior."

·ᗡ

At first I thought we had arrived at a cemetery, but I was wrong. I saw a glass enclosure that has commemorative monuments. Large busts rest on olive and gold columns that are Greek in style. The glass is in places decorated with ornamental designs and engravings. Around every statue there is a walkway, and between the walkway and the statue is a groundcover of grass similar to Mondo grass. The walkway enables you to examine every statue. In the center of the enclosure is a round gold platform, about two inches off the ground, and it rotates. On it are the most prominent and celebrated individuals from our history, as recognized by these people. They are full-body bronze and gold statues. Next to each statue, a gold plaque provides information on the person: who he was, what his field of specialty was, and from which century and planet he came.

A distance further out from this round platform with statues, aligned in a square grid around it, is a row of busts. These busts are bronze and gold, and they stand on columns that are about one meter off the ground. In every corner there is a miniature full-body statue that rotates. Each of these also has a plaque with information on the person, and the plaques are illuminated by spotlights.

I saw hundreds of busts, of different sizes: busts of scientists, architects, musicians, artists, and writers. Further down the street there is another section of celebrated humans. It, too, reminded me on a cemetery. Further down there are other sections celebrating humans from the Empire of Atlantis; these include scientists, inventors, and talented individuals who made their existence possible. The memorials to our celebrated humans and achievers reflect how well these people have preserved our history. Much of our history and many of our great achievers have either gone unrecognized by us or been lost to us. What a relief to know that they have our history preserved – after all, it is with their help that we have written our history and that we have become civilized. Our progression throughout the ages has been under their watchful eyes. When you come to think of it, one generation of these humans has lived to see much of our history, if not all of it. As a part of the education curriculum, students study our history. These people are proud not just of our collective achievements, but of those individuals who have made a difference on this planet.

I saw Leonard da Vinci, Shakespeare, and Plato. I saw Stradivarius. I saw Mozart, Bach, and Strauss. Not only does Shakespeare have a bust dedicated to him, but he also has a full-body statue in the center display. Here he is depicted nude, with a book in his left hand and a chalk in his right hand. In front of him is a blackboard. With his right hand he is pointing to something written on the blackboard, and that something is the start of a new play.

Leonardo da Vinci also has a full-body statue dedicated to him in the center. He is holding an invention: a model airplane carved from wood. There are two or three other celebrated humans sitting on chairs, and one or two on lounges.

I felt fortunate to have had the opportunity to have seen all of this. How marvelous will it be for us to one day learn of our true history, and of individuals we celebrate today whom we know so little about. I remember saying to the crowd of people with me, who were telling me about individuals they celebrated, "One day I am going to write a book about all of this."

"Oh! That will be wonderful," one said.

"Whatever you write, we will print it for you," another said.

"How would you print it?" I asked.

"You give me ten books, and I'll print them for you. Imagine if you were to sell a book printed by us – it would be worth one quarter of your entire planet." He then turned to Guyd and suggested they show me how they print books.

Guyd agreed and said, "Let's go straight away." A group of around twenty people came with us there. I must admit, I loved the attention. It is every actor's dream.

26

A Book Printer

We went into a building, straight to the man in charge, to whom Guyd explained that I was from planet Terra. The man was so excited about meeting someone from planet Terra that he hugged me.

Guyd said to the man, "When he goes home, one day he's going to write a book about us, and then he's going to send us ten books." He turned to me and said, "I'll tell you how – I'll send someone to accompany you back here."

We went into a room, and in it were several printing machines, which catered to producing different quantities of books. We went to one of the printers. All I could really see was a table. The table was in front of a wall. A panel on the wall in front of the table looked like stainless steel, and it was approximately two by two meters in size. In the panel was a slot that opened to a specified size.

The man in charge took a dust jacket from a hardcover book and placed it on a grid on the table. He also placed the hardcover book on a grid near it. The grid adjusted to correspond with the size of the book. When he opened the book, the left-hand cover rested on the grid that kept the top pages of the book level. He made adjustments so that it was a perfect fit. When he pressed a button, a clear cover appeared on it, and then the grid with the book moved into the slot in the wall. In normal circumstances the slot closes. However, it was left open for me to see some aspects of the process. There were lights from above the book that penetrated it. The only indication that some process was occurring was a faint whispering sound. Several minutes later, the 400-page book was copied, and an identical hard cover copy of the book was produced, all printed and bound, with a dust jacket. Depending on how many books are ordered, books are produced by this process and are transported into a box; when the box is full, there is an automated process by which the box is closed and sealed. Then the box is shifted along the equivalent of a conveyor belt into a storage pile on a loading plate. When the load is full, it hovers away and a new loading plate appears.

What may interest us is that they publish books written by us humans that appeal to their population. In general, a great many of our books don't interest them. I know that one author interests them as much as he interests us: William Shakespeare. A book was printed in Australia in September 2011 and is creating a bit of a stir beyond Earth. By the end of May 2012, it had not only sold over a million copies, but hit the newspaper headlines on Atlantis. The latest word is that while the people find the book extremely philosophical, they find it confusing: confusing because they cannot believe that humans from Terra could have written it. They know that those involved in the book are not from Atlantis, or from another planet – the latter was an initial consideration. They have concluded that they must be the reincarnation of humans from a superior planet. They cannot figure it out; what is puzzling is that they cannot get the answers from the minds of those involved in the book. It has become a conundrum to them.

The people of Atlantis are surprised about where the book came from. They remember a young man who was with them for thirteen days. It took fifty years for a book to come out with his knowledge. How this book was delivered to Atlantis is a question no one can answer.

Word of mouth has caused the book to become a runaway bestseller, and it has gone beyond the borders of Atlantis and is presently selling elsewhere in the universe. No other book written in our history has achieved such a result on Atlantis or created such controversy. The day the book was delivered to the garage of the author, late in September 2011, was the day it was published in the same format on Atlantis. By mid October 2012, the book was still on the bestseller list.

27
Abductions

Not only do the Atlanteans collect books that interest them, sometimes they collect humans that interest them. These humans are taken to the mother ship primarily for experimental purposes. No one is ever harmed in any way. The Atlanteans collect from humans data that they consider relevant. This may be pertinent to their genetic engineering program. After retrieving data, the person is returned to Earth, with no recollection of the events that have transpired. Other civilizations also study us. Some study our ability to fight harmful viruses and infections. Whenever a new strain of bacteria or virus is "bred" by us on our planet, others out there, including the Atlanteans, are interested, not just in analyzing it, but in finding an antidote for it. Animals and human specimens are thus sought for as test subjects.

Those specimens from Earth who are taken on the mother ship for experimental purposes may spend up to four days there, in their own physical bodies. But they won't know where they are, and they will remain inside the research facility. They are not given a tour as I was. When they return to the surface, they will have no recollection of their experience. In the process of this research, the Atlanteans may find some preexisting medical condition in the human, in which case they will usually eradicate that medical condition. For instance, if they find that the test subject has cancer, they will rectify that condition. When this human is sent back home, he will believe that he has experienced a miracle. But there is no miracle. There have been several instances of this occurring throughout our history.

Other civilizations out there are also interested in our human DNA; instead of researching us directly, some will cooperate with the Atlanteans. Why they would cooperate is that the Atlantean Empire has a "claim" to this planet: the Atlanteans were first on this planet, and, officially, we are their "project." Sometimes the Atlanteans help other civilizations in their research projects by providing them with the relevant data they require. What this tells us is that we are only tenants on this planet, and that no one country can truly claim to have a stake in its soil. When the time comes, the Empire of Atlantis has the right to dictate the paths that we take, which is something some of us will welcome, just as some of us will not.

28

Reporters and Cameras

I felt that a great deal of attention was given to me while I was on the mother ship. Just as I had many questions of them, so the people I met had many questions of me. This was all overwhelming. Often I felt a little nervous. There were occasions that I was in the company of delegates of the community, which captured the attention of reporters. Apparently, reporters even exist in a society as advanced as theirs. Reporters are robots, and they are also photographers. They took many photos of me. Added to that, everywhere we went, ordinary people were taking photos of me.

I found their cameras particularly interesting. One of the reporters was happy to show me how his camera worked. He said, "This is a professional-grade camera; it is the best on the market, and it is unbreakable, even if it falls to the ground. Here, take a hold of it." The camera I was looking at was about eight by eleven inches in size and thin. It was so light, and it looked simple in design.

"Can you see the square feature on the right-hand side of the camera, which is approximately half an inch square in size?" This was three quarters of the way up the body of the camera. "This is the lens. Can you see a flat button on the side of it? This is the exposure button. Press it and see what happens." I pressed the button, and heard a "click" and a "zip" sound – which is similar to what you hear when you take a photo with a typical camera. "There is no flash unit built into this camera.

The camera is sensitive and works on automatic exposure. There are settings on the camera that allow for conditions of night or day. In either of these instances, the finished photo will always come out the same. If there is too much available light, the camera will adjust the sensitivity to suit the relevant lighting conditions."

What I found amazing about that model camera was that its entire back was a viewing screen, as was its front. This enabled the subject to see on the front screen what the photographer saw on the back screen.

It is hard to picture that this type of camera, which I found so easy to use, was the best on the market when I was there, when you consider that it had no technical aspects to it involving the manipulation of apertures and shutter speeds, for instance. The only thing you needed to do was to set it to day or to night, and then press the exposure button.

On each side of the camera were two handles with which to hold the camera. It was easy to press the exposure button with your index finger while holding steady the camera. As for its images, the camera model I witnessed could store thousands of high quality three-D images on a memory chip that was inserted in a compartment just before the shutter release button. To remove it, you merely had to press the back of it and it popped out. When I pressed the exposure release button to take a photograph, the button felt soft, as though it had air in it. It was sensitive to touch. This meant that you did not have to press the button with much pressure and put stress on your joints.

"When you see the finished three-D picture quality, you won't believe it. Most people don't use this grade of camera for personal use; they carry a miniature version, which operates in a similar way.

"This professional version, however, is capable of producing instant pictures. The print comes out of a compartment on the left-hand side of the camera. Can you see it? A side door opens and up to ten pictures at a time can come out. There are photo shops that develop photos for those who don't have this version. As a reporter, I have to have this type of camera because we need to print photos in the fastest possible time to meet deadlines."

They also have newspapers and magazines. What this tells us is that nothing goes out of fashion. People will always want to feel a book, magazine, or newspaper in their hands. That being the case, there is a vast difference between the quality of their journalism and our journalism. Just as we have newspapers and magazines that are dedicated to local news and world events, so they have newspapers and magazines dedicated to local news and world events – these world events involve us humans. They keep up to date with everything that is going on in our world, in all of our countries. They also carry most of our television channels. They have news channels dedicated to events going on in the world. We may consider that we have a great deal of channels; however, they carry thousands of channels, not just their own channels and ours, but channels from other civilizations and planets all over the universe. Their channels do not operate on our frequencies; they use different technology, which is why we cannot tune in on their programs at this present stage as they can tune in on ours. How fascinating would it be if we could!

29

Antarctica

Guyd often sat with me on my lounge, with something interesting to show me. There was so much for me to learn and know. One topic involved Antarctica. The following is a layman's version of what Guyd told me. He spoke of experiments they have been conducting. I have to admit, I am unable to reiterate most of what he said. One thing I do recall is that Guyd said, "The melted Antarctic ice is pure and clear, as though it has just come from a mountain spring, with all its elements intact. It is the purest water on this planet."

On screen was a scene depicting Antarctic ice above the sea floor. "Can you see the huge flow of water beneath the ice? This is a river of fresh water. The sweet water is to salt water what oil is to water. The river is extremely wide and deep, and it flows at an unbelievable pace. It is always on the move.

"The sweet water river flows to the bottom of Antarctica, and no one has been able to find its beginning or its end. On top of the river, you have a mix of cold water and ice forming, which constantly breaks and joins, like a man that cannot sleep at night, who twists and turns until he finds a comfortable position. That is how Antarctica exists. The first question that comes to mind concerns how the gravity of Antarctica, and the ice and salt water of the oceans, have not been able to cancel out the fresh water rivers – there must be a few of them.

"These sweet water rivers house incredible marine life, which is handy for us because from time to time we go there and harvest food that appears to grow on the seabed. We have found many species of vegetables growing there – of course, they look different from the ones you are familiar with, but they have similar properties.

"It puzzled us at first, but the very existence of this marine life is attributable to the sweet water. As you can imagine, it is impossible for these species of vegetation to exist in salt water.

"When we first saw the vegetative growth on the seabed, we surmised that there had to be a colony in the vicinity, established by a civilization from a different planet. We thought that in such an environment, this civilization would have had no need to venture out for many years. We thought that the people may have been as self-sufficient as we are, and that the food that grows naturally in the sweet water may have been a trace of their existence. Perhaps even of their past existence, if they no longer have a presence there. But we still haven't come to the end of the investigation, so we are unable to dismiss that theory altogether.

"As for the vegetative growth on the seabed of the fresh water rivers, in dismissing the theory that another civilization is responsible for seeding it, we can only conclude that the seeds originated from a lake that we know of, and have evolved in such a way that they can exist on the seabed as they presently exist.

"Our scientists don't know where the lake is, or where it is the rivers go to reach the lake. The main question that we cannot find the answer to is where that fresh water comes from. We have concluded that the rivers that flow under the salt water must be filling in spaces in the Antarctic and creating lakes of sweet water. We envisage that the lake I mentioned earlier is significant in size. We know that it is exposed to sunlight, and that it rains on its surface. Additionally, it has a huge whirlpool in it. The fact that we have seen pictures of it does not mean we know where it is. It appears to be a completely separate entity in the Antarctic.

"The lake would have evolved by a process of different forces, of hot and cold, and this would have created tremendous thunderstorms and rain in that area.

"The lake is warm, with vapor rising from it. Warm air rises and collides with cold air. Then collisions of negative and positive energy create thunderstorms and rain in the area. That rain then falls on Antarctic ice and freezes instantly.

"What is going on in the region is a rare occurrence that we have been able to discover, but it is not unique to this planet; it occurs all throughout the universe. One thing is certain: there could be hundreds of these rivers, under all of the oceans."

30
A Visit to a Recreation Area at the Lake

One day I set out with a small group to see a popular recreation area on the mother ship, which is located opposite the lake. The area is their equivalent of a national park, where all the species of wildlife are protected. I saw many wild animals inhabiting the area, including rabbits and antelopes. I saw small hills, varieties of trees, and even a rainforest. It is hard to believe that all this could exist in a flying saucer.

I was told that visiting this place is a popular pastime, for not only the kids, but also the adults. Kids are encouraged there so that they can learn about nature. Tuning in on nature begins from young.

On the rim of the park is a magnificent waterfall. Guyd said, "The overflow of water is pumped back into the lake by way of a large piping system."

I saw red salmon, which are driven by a primitive instinct that has never left the species, jump up against the flow of the waterfall in a desperate attempt to beat the unbeatable odds of reaching the top – a feat that no will or determination could achieve.

One of my companions said, "She will get there one day. One day she's going to find herself in the huge pipe that pumps this water back into the lake, which is where she wants to go. Slowly, slowly, in time, she will wind up back here again, and then we'll have a no-win situation because she'll again spend some time trying to get up the waterfall, which she can't do. Eventually she'll be sucked back into the piping system, and, along with the water, she's going to be spat out in the middle of the lake. And she'll swim around the lake, which will be unfamiliar to her, even though she's swum those waters many times before. These are not smart fish; they don't have much in the way of intelligence. Smart fish won't wind up in this cycle because, number one, they know what the waterfall is, and, number two, they are too big to fit through the wire mesh that prevents large fish from going over the waterfall. While the wire mesh, which never rusts, by the way, prevents bigger-sized fish from going over the waterfall, it does allow smaller fish, up to a length of ten inches. As you know, the smaller species does not have intelligence of the kind the larger species has."

Beside the steep slope of the waterfall are steps and lookout points from which to observe both the waterfall and nature. Many people on each side of the waterfall were staring from various lookout points. There is a four-foot high protection screen at each lookout point, and kids are not permitted there without their parents. I assumed that there must be antigravity technology around the lookout points to prevent someone from falling over the ledge.

At the bottom of the waterfall, people were fishing. "They are allowed to fish in this area because there are no large fish here. Having said this, had there been a larger fish – on occasions there are – he would not be stupid enough to get hooked on to a fishing line. The smaller fish certainly are a different story. You can catch them and take them home to eat. Fish are plentiful here; in a matter of half an hour you are capable of catching around sixty."

"No wonder I can't see any fish shops around here," I commented. Everyone laughed.

"We all know where we can fish, what fish we are allowed to catch, and what size we can accept. The same with lobsters. You have to measure them before you can put them in your sack."

I noticed that kids are kids, no matter where or of what intelligence. Some kids by a shallow pond were picking up rocks and tossing them at fish. They were amusing themselves with this activity. Parents reprimanded their children for this behavior when they caught them, and warned them not to do that again.

In this area, the pond is only around six feet deep, and it is a popular recreation spot for children. Many children here were fishing with fishing lines. Yet some persisted in being cheeky and were tossing rocks at the fish. I remember watching one boy with interest. A nearby adult caught him and screamed out that he should stop doing it; then she notified his parents. How she notified them was by screaming, "Watch out, your child is killing fish with rocks. Maybe he'll find a big one, and maybe she'll throw the rock back at your child. I don't think you'd like that, so you had better do something about it." This was all telepathic. The mother then yelled at her kid, threatening to pack up and go home if she caught him do it again. Just like a typical kid, he stopped doing it for the moment, but then when his mother and the other adult looked away, he went back to his old tricks. His mother caught him at it again and stayed true to her word, for, she packed up and left, but not before giving him a gentle slap on the backside.

One of my companions said, "There are occasions that you may find a smart fish stuck here, like a dolphin. And if she gets a rock tossed on her head, she will find the culprit. She will peer out from under the water; at the same time she will read everyone's mind. Once she figures out who was responsible, she will grab that rock with her mouth. She will stand upright and start making a racket. Then she's going to spin in a circle to build up enough momentum to propel the rock, and the rock she throws will hit the target with one hundred percent accuracy. Dolphins enjoy playing with kids, but they do not want to be hit on the head! If one is, she will seek out revenge. Dolphins are quick to turn to the negative side over such things and seek out revenge on what they perceive to be their antagonist. In other words, they are temperamental and touchy.

"There was one occasion that I remember this happening. A child hit a dolphin on the head with a rock. When the dolphin saw the child responsible at the water's edge, in one swift action she grabbed a hold of him by the pants and pulled him into the water. The kid was screaming, which caused his parents to rush over to see what was happening to their child. The father questioned the dolphin on her behavior. Upon learning what his child did, the father apologized on behalf of his son and said it would never happen again, because he was going to punish his child for such stupid behavior. On this basis, he requested that the dolphin bring his child back to shore, which the dolphin did, but not before the parent had to do a bit of negotiating with the dolphin.

"Some time later you could see the same boy riding the very same dolphin, as though nothing had happened. As easily as the dolphin took offence, did the dolphin forget the offence."

One thing that I noticed was that rabbits had not lost their instinct to run for cover at the sight of a human. It appeared that, unlike many other species of wildlife, they had not developed a trust of humans. Guyd said, "Indeed, they are wary of humans, and for this reason do not interact with them. It will take a long time for them to interact with humans in the way that many of the other animals do, and it is little wonder why. If you look closely at their history, they have not had a lot of reason to trust humans. They have a good memory of their history, which is one of being killed, skinned, and sold to the public. Their intelligence is aware of this; while they may be intelligent, they are not intelligent enough to trust humans. They simply run. In their cycle of life here on the mother ship, they don't cause any problems; they only eat vegetables that grow naturally in the wild, and they shy away from human contact."

I saw a great deal of stags. I have to admit that I was a little scared of them. They have their freedom on the mother ship, without fear of predators, because not any exist there.

While Guyd constantly reminded me that there are no dangerous species on the mother ship, I did have a few hair-raising moments. One of those moments included confronting a snake. This was a harmless variety, but my instincts got the better of me. Guyd said, "If you were to camp out here, in the wilderness, you would not have to cover yourself with a blanket to protect yourself from nuisance creatures, or worry about being bitten by something poisonous, as you have to on the surface of the planet."

It is interesting to think that these people have camping sites in the mother ship, and that they frequently go bushwalking and camping, complete with a tent and a campfire.

Also interesting is that small monkeys live in the national park. What I didn't expect to find was that the monkeys have not developed a friendship with man.

Monkeys fear humans just as the rabbit species fears them, for good reason. Guyd said, "Kids have ensured a divide between the monkey and the human, because of the bad habits of the children. For instance, children have made it a pastime to throw rocks at monkeys and chase them. Because of the behavior of children toward monkeys, monkeys have grown wary of humans and have never had a chance to become close to, or trust, humans in the way that dolphins and birds have."

I saw an area of coconut trees, and on the coconut trees were monkeys. I was watching the way that the monkeys handled coconuts: first, they threw them from the tops of coconut palms down onto the rocks below. Then they climbed down the trees to retrieve them. It fascinated me to see a monkey sitting on rocks, using a sharp rock as a tool to beat open a coconut. With persistence, he managed to make a hole in it. The first thing he did was reward himself with the coconut milk. Then he used his strength to prize open the coconut and eat it.

Someone from the group said, "Kids enjoy eating coconuts too; the problem is that to get a coconut they often steal from monkeys. When they see a monkey as you just did, kids gang up on him. Of course, the monkey flees, usually up a coconut tree – often all he has left of his hard work is a piece of coconut in his mouth. The kids then help themselves to the rest."

"Now I understand why monkeys don't like humans, particularly kids."

"That's not all, once a monkey is forced to leave his prize behind and retreat up a tree, he retaliates against the kids by throwing coconuts down at them. And do coconuts hurt when they hit you! So this is what goes on between monkeys and humans. When you cannot find your kid down there by the water, you know where to look for him. He'll be here, annoying the monkeys. You can also bet that when you hear monkeys squealing, there are kids menacing them. Naturally, the monkeys reciprocate this behavior.

"Monkeys have not evolved to the extent that they can speak telepathically. They are in a different league to the species that can. You simply cannot talk to a monkey. Monkeys are too stupid, and have not evolved." I found this interesting in view of Darwin's theory of evolution. One's natural inclination would be to suspect that man's so-called ancestor would have something in common with man, and be the closest species to man. On the contrary, the monkey has as much in common with man as the rabbit has. You can say little about Darwin's theory, without either giggling to death or dying of embarrassment.

31

A Visit to Two Small Retail Shops

Guyd and I were at the shopping centre, and we happened to pass by a fruit shop. Because food is predominantly packaged as tablets, one could be inclined to think that there is no place for a fruit shop in the marketplace. Yet there are small businesses in the mother ship that sell food products which are not in the form of a tablet, one of which is the local fruit shop. Outside the fruit shop was a beautiful display of apples, on a gold pedestal that was about three feet high. On it were the most beautiful red apples. Each apple was large, and had a tiny stem and a lovely green leaf. The apple, I thought, was a piece of art on its own. Even though it looked tempting to eat, its artistic quality had captured my attention.

I walked into the store and approached the sales attendant, who was standing behind a counter, "Can I buy an apple; one of those out there?"

"Yes, you can go and get it."

Innocently, I tried to remove one, and the strangest thing happened. My hand went right through the apple. Additionally, all of the apples disappeared. While the pedestal was real, the apples were a holographic image. That brought a smile to the salesman's face, and even to Guyd's. Almost immediately, the apples reappeared. I kept doing it, and noticed that the shop attendant went from shaking his head to laughing aloud.

With a smile on my face but fear in my stomach, I said, "I don't know how you can eat an apple like this. I haven't got a clue, because I can't even get a hold of one." I was utterly befuddled.

"I'm sorry, just a minute," the salesman said and then walked over to a shelf and removed an apple; he polished it with a clean cloth that he had on his arm. After he placed it in a carry bag, he handed it to me. The apple was identical to those on display outside his shop.

I accepted the bag, and both Guyd and the salesman started laughing again. The salesman said, "Put your finger here, and then you can go." He was still laughing, as was Guyd. I put my finger on a device built into the table, which had a light emanating from it. I had never heard of eftpos technology at the time, so it was impressive to me.

Guyd and the salesman couldn't stop laughing, particularly when I said, "Anything else I need to do?" I couldn't believe that one fingerprint was enough to settle the transaction.

"Now you two can go and have fun. I can recommend the ice cream people two doors up. They sell very good ice cream."

"Ice cream! You mean, I will try to take an ice cream in my hand and it will disappear like your apples!"

Having composed himself, the salesman said, "Most likely it will do that. Many shops have such displays to attract customers."

I shook my bag and said, "What about my apple here? I can't wait to have a bite of it. But can you honestly tell me that when I bite it, it won't explode?"

"What do you mean, 'explode'?"

"Like a bomb!"

He put his hands on his head and laughed uproariously. Guyd was tapping one hand on the counter and laughing in the same way. The salesman said, "We don't sell apples that explode after you bite them. That would kill you. We don't sell things like that, things that explode. You go ahead and enjoy your apple. I guarantee it will be an apple and not a bomb."

"What will happen if I take one of those apples out of your presentation display out there and put my apple in its place?"

"Number one, you can't do it because the apple will always reappear in one twenty-fifth of a second. You would not have time to replace it. Besides, it's not physical matter. But if you could do it, then it will most definitely explode and blow your hand off, because you would be putting matter into non matter. Ah . . . enough of this conversation! I have never had a customer like you. My stomach is starting to hurt, not to mention my mouth. I have never laughed like this in my entire life. Now, I'll escort you to the door and show you the ice cream shop. You should go and buy one to cool your mouth."

I looked at him with an air of seriousness and said, "You mean I can have a bite of that one!" I was referring to the holographic image on display outside the ice cream shop. He laughed uproariously again, and with great difficulty managed to say, "I would make sure I didn't put my tongue anywhere near that."

At the front of the ice cream shop, right near the entrance, was a thin pedestal, which displayed ice cream in a cone that was turning. Different flavors of ice cream were represented, with each flavor constantly changing color to another. You could not walk past a display like that and not be tempted to buy an ice cream.

I saw a kid pass by the shop and try to grab it, and it disappeared for a moment, just as the apple did for me earlier. The kid didn't find it amusing, but I did. He ended up staring at me because I was laughing at him. Guyd was laughing as well, although I believe we were laughing for different reasons – I was the object of Guyd's laughter. The kid just took off with his father, who was holding his hand. The father said to him, "No ice cream today."

Inside the ice cream shop was a man with his son. I overheard that the man was buying a one liter tin of ice cream to serve as dessert for some guests he was entertaining that night. My first impression was to question why anyone would buy ice cream that could melt by the time you arrived home, when you could obtain it in a tablet form.

In response to my thoughts, the saleswoman responded, "All of our ice cream is sold in tins, in bulk, and it never melts, not until you open the tin. From then on, it will melt if it is not kept refrigerated. As you can see, our ice cream is stored in tins, which are kept at room temperature on the shelf, and they can stay on the shelf in this way for twelve months. This is because the container has properties that keep the ice cream frozen. Yet the exterior of the container is normal to touch." She gave me one to hold.

"You must be new. How did you come here? Did you come from far away?" the boy asked.

"Not really, I was just fishing at the lake, a couple of miles from here. Then I took a ride on a dolphin at the top of the waterfall, and winded up somewhere. I got lost! But, luckily, I found this man here." I pointed to Guyd, and then started laughing.

Guyd grabbed my arm and said to the man, who didn't understand my humor but looked puzzled to say the least, "Don't listen to him, he is just playing games." Guyd burst into laughter as he said this, and couldn't stop laughing.

Guyd's laugh was infectious. I believe that the man and his son were laughing at Guyd's behavior, and not at my comment. The boy then said to his father, in a forthright way, "I want to ride on a dolphin at the top of the waterfall, just as he did."

The man turned to me with a serious face and asked, "Why did you have to tell him how you came here?"

"If you don't know, I won't tell you," I said.

This was a serious exchange. I could tell that he was not impressed with me. After paying for his ice cream, he quickly got a hold of his son and left the shop, with not a word to anyone. This showed me a little of his use of his negative side. He had virtually told me off. It didn't look as though he was impressed that I was playing games with him. This one had no sense of humor. This was an occasion that I experienced a human who allowed his negative side to prevail. I had spoken with hundreds of people from Atlantis, and not once had I experienced a negative response from them. Someone in his shoes should have had a sense of humor.

Guyd said, "Kids are not permitted to ride on dolphins in the lake near the waterfall, just in case one of the dolphins gets worked up. Sometimes a dolphin will get worked up when he's having too much fun. He starts spinning and accidentally tosses the kid over the waterfall. They have a lot of power when they propel an object and can easily toss a kid the wrong way. There are signs all over the place warning people and dolphins to keep clear of the waterfall. Also important is for them to keep clear of the wire barrier. In the past, dolphins used to play a dangerous game with kids, which is why it has been banned. Dolphins would toss kids, just as you would toss a football, into the wire barrier before the waterfall, resulting in the kid bouncing off it back to the dolphin, who would catch the kid with his mouth. To the dolphin, this was fun. From that point on, dolphins were banned from entering that section of the lake, hence the warning signs. Some young dolphins get too enthusiastic when they play with kids. Often they don't know what they are doing, and are not aware of the consequences. Another thing is that the kids cry when this happens, and young dolphins find it entertaining when kids cry; this is because they are confused and think that the kid is enjoying the game, until someone tells them otherwise."

I understood why the man was not impressed with my comments. Perhaps one cannot fault him for letting his paternal instinct override reason. Perhaps the problem was that his son would have given him a tough time from then on. Even so, the boy still appeared to have a bit of attitude, which is hardly surprising when you consider his father's attitude and behavior.

After he left, I asked the salesgirl, "Do you sell whiskey or alcohol here?"

"We don't sell alcohol in this shop."

"It looks to me that he's a little drunk."

"Most probably he is, but we don't usually drink alcohol, and it doesn't affect us anyway. But you're right; the boy's father looked as though he had some drinks."

I headed out, and Guyd followed, which prompted the salesgirl to ask, "Aren't you going to buy an ice cream?"

"I'd love to buy an ice cream, like this one here; it looks tasty, but someone said it was going to freeze my tongue if I try it." I was referring to the outside display.

My answer surprised her. She turned to Guyd and with a puzzled look said, "I don't understand."

He smiled and just brushed his hand through the air in a dismissive manner. He turned to me and said, "Let's go before you upset someone else."

32

A Tribute to the First Visitor from Terra

Guyd was taking me on a mystery trip. What he was up to, I had no idea; I only knew that we were heading back to the beach for a second visit. This time we traveled there in a shopping trolley. Once we arrived, Guyd said, "I have a small surprise for you." I could not imagine what that surprise would be. "Come with me and I'll show you something." We stepped out of the shopping trolley and walked a short distance. I was shocked when we neared the beach. I saw over one thousand people there, and all were clapping and staring at me. What I found even stranger was to see dolphins balanced upright in the water, also clapping.

My first instinct was to say, "What is wrong with these people! Are they crazy or something?"

"This is in your honor. Now they all know who you are."

At that moment, I noticed the same dolphin I encountered last time I was at the beach. He was amongst the other dolphins, only he had a sardine in his mouth. When I concentrated on him, I received a telepathic thank you from him. I was dumbfounded. I knew why he was thanking me: I had made him famous, and I was responsible for his regular one-ton supply of sardines being dropped off in the water, which he could share with his friends.

In no time, spectators surrounded me, as though I were a celebrity. A delegation of the most prominent members of the community, such as the mayor of the mother ship, began to take turns in shaking my hand. The mayor was not a robot. He was a member of the community, elected to serve in his position. We stood together for an official photo. All of this baffled me. I spent a while shaking hands and greeting people. Everyone had a question about our planet. For instance, one asked, "Are you a violent people or a friendly people?"

"We have people that are friendly, but we have people that are violent – like your coconut-throwing monkeys."

"Monkeys! Monkeys!" He looked at his friends around him and said to them, "Oh, god forbid if we have lots of them to deal with; we have enough monkeys here to deal with as it is!" Everyone was amused.

A woman took the opportunity to ask, "How do you dress?"

"Nothing special, very similar to you."

Girls in a group of young kids began to throw random questions at me.

"Do you have shopping trolleys?"

"No, we have cars that run on a combustion system."

"Oh, we hover. We don't even have to worry about driving anywhere."

"Oh, don't you have flying saucers to go out?"

"No. We don't have that one. We have planes."

"Are there nice boys there?"

"Heaps of boys, as many as you have fish in the water here."

"Are they nice looking?" The girls were giggling. Rather than answer such a question, I just smiled.

A girl who must have only been around twelve years old asked, "Are you married?"

"No, I wouldn't get married so early. I am too young."

"Young!" she exclaimed, with eyes wide open and brows up in surprise. "You could be my father." (When I was young, I always looked about twenty years older.)

"Well, since you put it that way, I could, but I won't."

My comment made them laugh and run away. It was obvious that they were having fun with me, and making conversation just for a kick. I did learn that no matter how intelligent kids are, kids are always going to be kids.

I was with a group of around fifty adults, and I asked them if they had any questions they may want me to answer. I was a little surprised that they had none. One responded, "No, we learn everything we need to learn about your planet in school, and we take the knowledge from your mind once we meet you. We already know your entire history; how we know all about you is by reading your mind. You may not know it, but you have already told us."

One did say, "We find the way you came in here unusual. We normally go and hunt humans for the purpose of research. We also do this to humans from many other planets. Then we return them unharmed, but with a loss of memory of the incident. Some lose three or four days of their own time. But you just showed up with your complete energy. You left very little energy in your body back in hospital."

"Well, I didn't know that it was possible until it happened; I just found myself on the ceiling looking down at my body."

"We always have people on your planet, robot guardians, who are looking for evolved humans – those who are unique in the sense that we can bring them here as we have brought you. We can do the same thing: we astral travel all the time. We want to find people who are advanced that can learn and know about us. We want to catch such a person and bring him here metaphysically, just as we brought you here. As you mentioned earlier, you are going to write a book one day. You are going to further the knowledge of your society. But you won't find people like you on Terra who can astral travel. You were caught because you left with a lot of energy, which makes your circumstance unique."

After a while, someone brought a pedestal. Guyd said, "Pop up on it."

"What for?"

"They want to take a picture of you, and you should address the people with a word of thanks."

Guyd helped me step up the platform. I had everyone's attention. Once I introduced myself, everyone clapped. On their faces were smiles. In the distance, I noticed my dolphin friend, so I once again said hello to him telepathically and smiled. This was when the weirdest thing happened. He spun around and tossed a sardine at me. My instinct was to catch it rather than be hit with it. A photographer captured the moment with his camera. Everyone was amused at how swift I was at catching it. Indeed, it was a hilarious moment. Amid everyone's laughter, I just waved with two hands. Then I did a Shakespeare bow and blew a kiss to everyone. Everyone yelled and screamed.

"Turn back and have a look," Guyd said.

I turned back, and around six feet from me was a bronze statue, of me! It was on the edge of the walkway. I cannot explain what a shock, not to mention what an honor, that was. I could not understand why it was bestowed upon me. The statue depicted me with a smile. I was standing with my right hand in a pocket and my left thumb in another pocket. To create the statue they had relied on a photo of me that had been taken earlier by the reporter.

At the base of the statue was an eight by twelve inch plaque. Written was the following:

Jack Lord
(The date)
Our First Guest from Planet Terra

I was the first visitor from Earth on Atlantis. Even Plato did not have this honor. Plato's story is a different one from mine.

I was in shock. I stepped down from the pedestal and began to shake hands with everyone. One woman with two children by her side said, "I knew that you are not one of us from here, but are from the outside, from planet Terra." Everyone treated me as a celebrity. To say that I was well received is an understatement.

A representative of the Chamber of Commerce introduced himself to me and then invited me to attend a dinner being hosted by the Chamber of Commerce. This event was scheduled for the following day. He said that there was going to be around one thousand guests in attendance, and it would give me the opportunity to meet some more people, especially from the business community. I gladly accepted the invitation.

Eventually I said goodbye to everyone and left with Guyd.

・ノ

The next day I attended the business dinner, which was a black tie event. I wore a black suit, white shirt, and red bow tie. Guyd was present for the event and we sat together at the main table. Beside me was the President of the Chamber of Commerce – the same gentleman who had invited me. In attendance were both men and women. In one corner of the room were musicians and a small dance floor. Once the evening officially commenced, the President stood up and said audibly (which I found out later was for my benefit), "This is Mr. Jack Lord from planet Terra. He only has a few days before he leaves us, so we decided to bring him here to meet you all."

"It is a very nice thing that you did," someone shouted. "We're so pleased to see you, and don't forget to come back again."

The President had some additional words to say about me, before introducing me to speak.

With his prompting, I stood and said, "Thank you for your kind words. You people have been extraordinary human beings."

Someone from the audience interrupted me by laughing aloud. Another one yelled out, "Half of the population are robots!" He, too, laughed aloud, as did the audience. I found his comment hilarious.

After my short speech, the President lifted a glass of champagne and said, "Let's drink to it."

The one who made the comment about the robots said, "I'll drink to it." Another, "Me too." Then I heard a repeat of those words echo throughout the room. I heard the tapping of glasses, and joined in as all drank their wine to the bottom of the glass.

The courses were served. After dessert, there was about half an hour of dancing, which I did not participate in. I spent the evening talking to people, and was as interested in finding out about their way of life as they were about mine. I learned a great deal about business and manufacturing. You could write dozens of books on what I learned.

They were curious to hear what I had to say about planet Terra. I remember saying, "I don't have much that is good to say about a world that has wars, or even about the country I grew up in where there is misery, suppression, as well as sickness."

One person was a little confused and asked, "What is misery?"

"You're very lucky not to know what it means, let alone to experience it. Just imagine if you were on a mountain, in fifteen feet of snow, with nothing around you but one little shelter. You have nothing to eat, and your parents only make babies. Imagine that you have no shoes, so not to be barefoot you have to make your own shoes from the skin of pork, and then imagine that you have to walk one hundred and fifty meters every day, in those shoes, through that snow, to fetch water. All you can do is pray god that spring and summer hurries up and comes."

They all talked among themselves. "That's a funny, funny life," someone said.

"How can humans permit such things?" another person asked.

They genuinely couldn't believe it. It was so alien to them. I don't think that these people are capable of ever handling living in our corrupt world. This is why they have cloistered themselves in their little paradise.

"You are the luckiest people in this universe; god forbid anyone who lives on planet Terra," I said. After a pause I added, "It is a fitting name. When I think about planet Terra, I only think about terror. All my life I have lived in terror."

They laughed, but they gained an idea of what living on planet Terra means.

One of the comments about Terra was that we were too behind in our technology: that we should have been more advanced than we were. (We should remember that this was the year 1959, if I remember correctly, and there were no home computers and no mobile phones. In the last fifty years, our technological advancement has been unprecedented, and it partly has to do with these people and their influence.)

"You are a very intelligent person," was a comment someone made to me.

My response to this was, "I'll take your comment with a grain of salt, especially in view of what you have done here – for instance, how you use clean energy. As a young man, I used to wonder how it is we can solve our energy problem and produce energy that is safe and clean. Life took me in different directions, and my interest in the field of physics was not what fate had in mind for me."

After around two hours, I said to some of the committee members, "I regret to say that I am very tired; all this has been so overwhelming. I want to register everything that I have learned so that when I return to the surface I can write a book about it. Ultimately, I am still only a human, and my brain doesn't work as yours does. Once again, I thank everyone in this place. I hope that one day we all see one another again."

One of the men said, "When you write your book, we will try to get a copy of it and put it in the library."

Everyone shook hands with me as I left. This process alone took around half an hour.

33
Guyd Kicks the Bucket / My Legacy

I found Guyd's cheeks funny in that he had the brightest red cheeks ever. One never thinks of a robot as having red cheeks, particularly not that red. His cheeks were inspiration for me to one day say to him, "You're going to have a heart attack. Go and look at yourself in the mirror."

"You're a funny, funny young man. How do you think of these things?"

"Because if a human has red cheeks or a red face, it means that it wouldn't be long before he kicks the bucket."

Guyd was stunned for a moment, and then he asked, "What does 'kick the bucket' mean?"

"We humans are biological, and we depend on our breathing system. We are not full of metal like you! For example. You take a bucket. Maybe elevate it a foot off the floor. Find a strong rope and on one end make a noose. Then tie the rope on something higher up on the ceiling. Put that noose on your neck, step onto the bucket, and then kick the bucket out from under you. What will happen is that your neck is going to go 'kkkrrrkkk' in the noose, and then you're dead! If nobody is around to cut the noose within twenty seconds of your having kicked the bucket, you can be assured that bugs and maggots are going to feast on your body when you are six foot under and dead and buried."

His perplexed look intensified as he asked, "What is 'six foot under'?"

"If you dig a six foot hole with a shovel, and then if I cut your head off and dismantle you to pieces, then I will bury you in that six foot hole. But because you're a robot and you won't stink when you rot, I could bury you in a three foot hole."

"Oh, that's gruesome! I shouldn't have asked you. Tell me, though, what is a 'bucket'?"

"A bucket is what you use to clean the floor. You fill half of it up with water and a little soap or disinfectant." With my fingers, I drew an outline of a bucket in the air. "Then you need to get a stick with a mop." I also drew an outline of it in the air with my fingers. "The stick has to have strings coming off the end of it. Then you shove the mop in the bucket, and then you squeeze a lever on the side of the bucket with your foot as you pull the mop through two squeezing rollers. These rollers will drain most of the water from the mop. Then you put the wet mop on the floor and rub it. That will clean up all the dust and make the floor wet, which you should not walk on until it is dry."

"What is dust?" he asked.

"You know, when you smoke a cigarette, and the part that has burned out falls on the floor." With my fingers, I gave him an idea of the size of a cigarette.

He interrupted me and said, "I know about that from *The Three Stooges*. But they have a big one, which is thick." He used his fingers to illustrate the size.

"Oh, you've seen the rich man's cigar – that's Havana Cigar. That one will give you a lot of rubbish. Rich people smoke those big ones for half a day."

"We don't have such luxuries."

"Just as well, because that will make you feel as though you are drunk, only it will make you drunk for longer; added to that, the inhalation will go into your lungs and one day you are going to kick the bucket."

"You know something," he said, pointing his finger at me. "I have a good friend, and I'm going to make this bucket. You make a sketch of how it looks for me on a piece of paper, and I'm going to go in the delivery center when nobody is around, and I'm going to make a noose – like the old fashioned fishing trawlers use – you know, when they fish they have a special gadget they turn when the fish gets hooked on a net. I saw it on television. It is very primitive, but effective. When my friend comes in, I am going to ask him to stand on the bucket. But first I'm going to show him how to wash the floor. That will make him so surprised because we don't have dust, but I'll convince him that there is dust. Then I am going to put the noose around his neck. Then I'm going to tell him to kick the bucket. And that's what you mean by 'kick the bucket'!"

"Yes; but because we have biological bodies, we die quickly because the noose breaks the neck and stops blood from circulating in the brain. If that happens to him, then he'll go to sleep, and he'll never wake up."

After a moment of reflection, Guyd smiled and went, "Ha! Ha! Ha! Ha! He will never trust me again."

・ᴗ

The next morning Guyd came into my room and started shaking me in bed. "Wake up! Wake up!" He was laughing hysterically.

"What's the matter with you? Have you lost your marbles? You scared me to death." I looked at the clock and then added, "It's 8 o'clock!"

Guyd made himself comfortable by sitting on the side of my bed. The bed sunk considerably under his weight.

"I did it to my friend. Exactly what you told me. Remember? I made a bucket myself in the workshop, according to the dimensions you gave me. I also made a mop. I found an aluminium stick and I made around thirty holes in the bottom of it, and then I put small ropes in the holes and tied them up. Then I cut the bottom so that they were straight. My friend was at first very cautious, very suspicious. When I gave him the mop to mop up the dust, he stopped before he ever started and said, 'We don't have dust here. Why should I mop here?'

"I said, 'Sometimes fish come in here and leave dust that they drag from the bottom of the lake. So we are not exactly immune from dust.'

"He looked at me and put his small finger in his mouth to bite his nail like I do, but then he realized that he doesn't have a nail. You gave me that habit, because I have inadvertently started imitating you."

"I only do it when I am deep in thought," I said.

"He began to mop, exactly like I showed him. He mopped one corner two or three times, and then he tossed the mop to the ground and said, 'Ah, you're playing games with me!'

"So I said, 'OK, if you don't want to mop then you can do this. This is much easier. Let's see if you can do it.'

"He did not have much water in the bucket, so I emptied it into a container and said, 'You can do this one for me. This is something Jack explained to me, and humans like him don't live 1500 years. See how smart humans are!'

"I put the bucket in position in a corner where I had the noose already prepared. He asked me what the noose was, to which I answered, 'Don't worry, it's just an exercise thing. I'll put the bucket upside down and you step on it now – you don't have to worry about dust.'

"As soon as he stepped on it, I put the noose on him and said, 'This is to hold you so that you don't fall down.'

"My friend found the whole exercise suspicious. I said, 'Now you kick the bucket away from you.'

"He thought for a second and then began to panic. I said, 'Pipe down; you've got nothing to worry about.' It took me a few words to assure him that it was just a joke. Then I said, 'Now, I'll hold your hand just so that you feel comfortable, and know that nothing bad is going to happen – that I wouldn't do anything bad to my best friend.' Before I finished my sentence, I kicked the bucket away from under him. He started screaming and holding his arms out as though he were about to fly. When I saw that he was distressed, I got the bucket and put it back under his feet.

"He couldn't wait for me to loosen the noose; he somehow did it himself and tossed the noose like this." Guyd imitated his friend. "The noose unexpectedly hit him on the head in the process, and because he was still on the bucket, he lost control and fell on his backside. I started laughing, and I couldn't stop. You could see the red mark around his neck where his blood had stopped circulating. He even said that he felt pain. (This is because they have the same nerves, enzymes, and so on, that we have.)

"He said, 'You can laugh, but I will never listen to you again.'

"I said, 'I just wanted to show you how Earthlings kill themselves when they don't want to live anymore.'

"He got lost in the whole process for a moment and couldn't understand why Earthlings would want to do that to themselves. Then he said, 'By the way, how did you find out about this stupid joke of yours?'

"I said, 'Jack told me.'

"He said, 'Ah . . . you and Jack!' He waved his hands in the air and left. He walked a few paces, looked back at me, and then started running. And believe it or not, I looked for him afterward, but he wouldn't come near me. Sometimes he hides from me. If he can't hide, he turns his head and looks away.

"I always wanted to know what it means to kick the bucket. It's primitive, but it sure has substance." Guyd was laughing all through this conversation, and could not stop.

"Not for me," I said.

"No, I wouldn't laugh at you. And I don't ever want you to kick the bucket."

꩜

The following day, Guyd tried to do the same thing to someone else. He told me about his adventure the next day, when he woke me up at 10 o'clock in the morning. He said, "This other person threatened to disassemble me." Guyd just could not stop laughing. He said he found the whole thing to be so stupid and yet so funny.

"Apparently, the message about what I did to my friend got around. He has been doing nothing but passing on the message of how the Earthlings that don't want to live anymore kill themselves. He has been telling everyone of how I put him through that ordeal and nearly killed him, and of how scary it was for him. He warned them not to put themselves in the same circumstance that he put himself in. He also warned them to keep a distance from me. He didn't take long to warn everyone, telepathically.

"Now, no one looks at me, and no one talks to me. They all look away, but if they do happen to look at me, it is with crooked eyes. And then they turn their heads away and go about their business."

꩜

One day I was sitting with Guyd on a park bench, watching kids play with what looked like a soccer ball. The disorganized way they were playing with it suggested that they had no idea of how to play the game of soccer. For, they were playing in a haphazard way, with no rules whatsoever. What they were doing was kicking the ball around, and it made no sense. Whoever kicked the ball first got a point. There were no teams. I couldn't believe that such an intelligent society didn't know how to play soccer.

"These people don't watch sport from your planet. They are too preoccupied with intellectual things. But the kids like to kick a ball around," Guyd said.

"But nobody knows what they are doing! What they are doing looks stupid to me."

We were at the park that was right next to the beach. I could see a few dolphins in the water. That was when I was struck with an idea. I asked Guyd to tell those dolphins that I was going to teach them a new game. When I had their attention, I told them that I was going to toss a ball to one, and he had to make sure that he didn't let it fall in the water. He was to keep it in the air, and then pass it on to the dolphin on his right. Once the ball got to the last dolphin, he had to hit the target with the ball. A monument, which was nearby in the park, was selected as the target. If his aim was right and he hit it with the ball, then he would get a prize – what else but a fish! All were eager to play.

I asked the kids playing nearby if I could borrow a ball so that I could teach the dolphins a new game. I told the kids that later on I would teach them a game called soccer.

There were around ten boys, and they were around twelve years of age or younger. They were interested in the idea. Even Guyd was interested. He said, "It's a good idea for them to have some physical exercise and get their minds off computers for a while."

At that moment, an ice cream van neared us and stopped. The boys jumped up in excitement and ran off all at once. Shortly after, they came back, each with an ice cream cone. One of the kids said, "Here you are. Take one." He did the same with Guyd. We gladly accepted the ice creams. I thought that that was a courteous gesture, which surprised me because neither of us had asked for an ice cream.

I said to the boys, "First we eat; then we play." So there we were. The boys all sat on either the grass or the sand. Once we finished the ice creams, I commanded the attention of the boys and asked them to sit in an orderly way with their legs crossed.

I was telling them of the virtues of the dolphin. "You see, the dolphin is a very intelligent fish. If you happen to be drowning, he's going to save your life."

A boy interrupted me by blurting out, "Yeah . . . sometimes they swing a tennis ball and hit you with it, which hurts."

"Well, there are ways to temper the dolphin's energy. For instance, we're going to get the dolphins to play with this soft ball." I had the attention of the boys, and

I had the attention of the dolphins; I even had the attention of my friend, Guyd. I asked Guyd to call another four dolphins over, and have them form a circle. Once the dolphins were ready, I said to them, "I'm going to toss you a ball, and everyone has to pass it around to the next dolphin in the circle. The last one who gets it has a right to swing the ball and hit it at the monument. If you hit the target, you'll get a fish." By this stage Guyd had returned with some fish.

I threw the ball to one dolphin, who passed it to the dolphin on his right. When each dolphin caught the ball, he did some clever tricks with the ball, such as spinning it on his nose and bouncing it, before he passed it on. The last dolphin propelled the ball up in the air like a tennis serve, and then with his nose he whacked the ball, which hit the monument. Of course, he received his prize. The game continued in this way. What impressed me most about the dolphins was that they did not try to steal the ball from anyone else. They were disciplined, and kept their position in the circle.

The kids found this so interesting to watch. Everyone on the beach felt the same and gathered around to watch. All was going well, until one dolphin missed his aim and did not hit the monument. A boy of around eight years, who was on the beach with his mother, stood up and said, "What kind of stupid fish are you?"

When the ball was thrown back to the dolphins, that particular dolphin became undisciplined and grabbed the ball for himself. He flung it in the air and used his nose as a bat to propel the ball to the kid – it hit him on the face. The kid ran to his mother, dropped to his knees, and started crying. The mother said to the dolphin, "You are a stupid fish. You broke the rule."

With a quick wit, the dolphin said, "You're not a good mother, because your son insulted me and everyone here who is playing ball." All the dolphins then sided with him and agreed that the kid should not have insulted them. All the spectators aired their opinion, and it was in support of the dolphin.

The mother showed her negative side with a touch of anger. She took her child by the hand, and with her other hand she tossed her sun hat on her head and her handbag over her shoulder, before walking off. Everyone laughed and clapped at her. Even the dolphins clapped.

The dolphins loved playing that game – as long as they got a fish. In no time, many other dolphins noticed them and took interest; they, too, wanted to participate in the game. Now there were around twenty dolphins in the group. Even with this larger number of dolphins, they remained disciplined. They took their turns, and they certainly did not cheat.

More boys came and joined us, until there were twenty. With twenty dolphins and twenty boys, I taught them a new game. I introduced beach volleyball, without a net at first. There was a team of twenty boys versus a team of twenty dolphins. They had never played a structured game like that. Both the boys and the dolphins loved the activity. Later on in my stay, I saw adults playing, this time with a net; it did not take them long to create one. I told them to find out the rules from Terra. I taught them how to play in teams. I also introduced water polo to the dolphins. Nets were specially made for them. You should have seen how happy they were. Until then, they did not realize how boring their lives were, so they welcomed these new activities. Dolphins played water polo in teams against fellow dolphins. Dolphins played volleyball against humans. Humans even played against each other.

I told them to organize uniforms for their teams. As far as the dolphins were concerned, I went to a florist, hoping to find something that the dolphins could use that was symbolic of a uniform. I had the florist make an artificial rose that the dolphin could tie up under his chin, on his own. Each team of dolphins was given a particular color. The dolphins were so excited and proud that they even shook my hand. They practically came up to the sand to do this. Each dolphin kept his. He would hide it somewhere under a rock on the floor of the sea, and when it was time to play, he would retrieve it and put it on himself. The way it was designed, it did not fall off them once they tied it on. Dolphins are cleverer than we can imagine. I believe that they are more intelligent than many of us humans here on Earth are.

The next thing I did was teach the boys how to play soccer. Goal posts were made, along with nets. They too had uniforms made, and formed teams. Beach volleyball, water polo, and soccer were a legacy of my visit to the mother ship.

34

A Visit to the Museum

I boarded the equivalent of a bus, which is used on excursions. It is shaped like a bus; it hovers about three feet from the ground; and it can seat fifteen people at one time. It also comes with a driver. Because it travels slowly, it does not travel with the rest of the traffic. Two were used on that day to transport a group of us to the museum. All of those in the group I was with were visitors on the mother ship, and they had come from different planets. What surprised me most was that every one of them spoke English.

The street leading to the museum is so spectacular. It has many large buildings that are Roman and Greek in style. Frankly, I saw no modern buildings, as we would define modern today. I saw nothing futuristic. Everything they have is classical.

Once we arrived, Guyd left. I found myself in front of large entry doors that were open. The doors of the museum are gold, with ornamental olive engravings. Two women, appropriately dressed, came to the door, whereupon one said, "Ladies and Gentlemen, if you can follow us." She spoke audibly, and not telepathically.

We stepped inside and the same woman said, "We are going to split up into two groups. I will take this half of you one way, and my assistant will take that half of you that way. We will try to answer any questions you may have during this tour."

It is so important for us to know that the museum has a record not just of the achievements we have made throughout the course of our history on Earth, but of the people who have made those achievements possible. There are statues of them, and each statue has a short narrative on the person and his contribution to our history. What this illustrates is that the Atlantean Empire has been with us during our evolutionary process thus far, and has recorded all of our achievements, in all the relevant fields, many of which have been lost to us, particularly those in our ancient history.

The tour guide was articulate, and knew her subjects well. From time to time she would ask us if we had any questions.

Some residents of the mother ship were visiting the museum, and whenever they greeted us it was always with a smile. This made me think, "Smiling is important to these people."

The tour guide noted my comment and said, "You are correct. That is the positive side; it's a vital part of humans, which enables us to live in peace. Rather than use negative energy, such as in wars and fights, we learn how to smile and enjoy the life placed before us. This is why our life span is up to 1500 years – more in some cases. So let's continue with the tour."

I saw an image of Leonardo da Vinci. He is portrayed in his elder years, and he sports a short beard. There is a statue of Leonardo's Mona Lisa, based on Leonardo's portrait. Her hands are resting on a column. Shakespeare is also there, not only with a full-body statue, but with a bust. He is depicted with a walking stick, at around fifty years of age. (We have no idea of how Shakespeare looks, and the accepted image of him would have him turning in his grave. Fortunately, we will have the opportunity to see a sketch of him in another book, which is dedicated to his true life story.) Plato is also there. There is a statue of him slightly bending over a table. He is pointing a finger to a location on a map. It should be of no surprise to learn that it is to the location of the sunken island of Atlantis that he is pointing. Plato appears to be in his late forties or early fifties, and he too sports a beard. Plato's appearance is interesting, but this is dealt with in another book.

As well as respected humans from our past, we saw rotten humans. The Atlanteans have a section reserved for those singled out as having contributed to the decline of the society, or having contributed to society in a negative way. These included dictators and merciless killers. I saw Lenin and Stalin. I also saw Tsar Nicholas of Russia. For those who find themselves in this category, there is nothing to feel proud of.

Time seemed to pass so quickly, perhaps this was because everything was so interesting. It took us another hour to regroup back at the main entrance.

One tour guide stepped into the center of the group and asked, "Before you continue with the rest of your tour, which will take another two hours, would you all like to have a small break at a restaurant where you can have some refreshments?" Everyone agreed.

35

Lunch at a Restaurant

We took a short walk across the street to the restaurant, which was French in style. A man greeted us; he appeared to be expecting us. He ushered us to our seats. Those of us from the first group sat together at a large banquet table. Those from the second group sat at another. The table was set with fine cutlery and china. On the table was a delicate lace tablecloth. The chairs were in Louis XIV style furniture, with olive cushions.

A male drink waiter asked us what we wished to drink. His uniform was a white shirt, black bow tie, and black apron. We were given menus, from which I ordered a dish of Lobster Mornay. I always seemed to order this dish. Everyone at the table waited for me to order, and then ordered the same, along with French wine.

We spent the next forty minutes or so dining. In this time someone asked me, "Where do you come from?"

"From planet Terra."

Someone else turned to me and asked with an air of seriousness, "How did you get here?"

"I flew here."

"Oh, how do you fly?"

"I have my own flying time-machine."

Everyone laughed. The man who asked the stupid question also laughed, but he didn't ask me another question. Many of those in my group looked stupid, especially the one who asked the stupid question. Where he was from, I didn't care to know.

· ᴗ

After the dishes were cleared, fifteen women, all wearing a uniform of black skirts to their knees, white shirts, and black bow ties, entered the room in a procession. Each was carrying a white porcelain dish filled with blue fog, and on the left of the dish was a towel. They stood behind us and waited for a waiter to pull our chairs out from under the table. This enabled a waiter to pull out, from under each chair, a small table that had two legs. You had to spread your feet a little or move your legs to the side for them to pull out the table, which was only ten inches wide. The waitress carefully placed the porcelain dish on this. Then she positioned the towel on my left arm. I was surprised when she took a hold of my hands and carefully placed them in the blue fog. She even took the towel and wiped my hands. Then she removed the porcelain dish from the small table, and just touched the table, which made it fold away on its own. As she left, another waiter pushed my chair back under the table. I was amazed by the quality of service.

Once we were all seated back at the table, additional waiters and waitresses came in and took our orders for dessert. I chose vanilla ice cream with a sweet cherry – the cherry turned out to be as large as a small plum. Most of those in my party ordered the same. This was not the first time that I had eaten this flavor of ice cream while I was on the mother ship. I ordered an ice cream at least twice every day. The circumstances of my life up to then revolved around one central theme: poverty. I was lucky to have eaten a dinner that was filling once in three weeks. Therefore, I don't have to paint a picture of how I reacted to food once I got there. I ate anything I could get my hands on, particularly desserts. It was no wonder people often stared at me eat when I dined in public.

Some of those in my group ate with their hands; they were shoving food into their mouths as though they had never seen food such as this in their lives. Even though I ate as though I had never seen food in my life, I knew how to eat with manners. This is not something I was taught growing up; I had learned much from working at a cinema. I could not afford to pay for the admission ticket, so I managed to secure a position there just to watch movies free.

After I finished desert, three men who were unfamiliar to me entered the restaurant and approached me. The waiter pulled out my chair as I stood to greet them. I was under the impression they were from the restaurant, but that was not the case. One of them shook my hand and introduced himself. He said his name. It sounded Latin. Then he turned to the two men with him and said, "Well, this is Mr. Jack lord. He comes from planet Terra and is here on a visit." One of the two men then bowed his head slightly and shook my hand. He said, "That was a good choice of name." Then he told me his name. The third man shook my hand and introduced himself. Typically, I cannot recall any of their names. Then he said to me, "You are a nice-looking young man. Have you had a second look at our nice ladies serving you in this restaurant?" He had a smile under his tiny French mustache, which I thought was somewhat funny.

"Does the French mustache try to say something to a beautiful lady?" I asked.

He tapped me on the shoulder. "Ah, you read my mind."

"You do it all the time."

The first one said, "I've been to planet Terra; it is a beautiful planet."

"It is, unless the idiots there blow themselves up and kill us all with nuclear weapons."

He laughed. "Oh, I wouldn't worry about that. We can stop those sorts of things in a matter of minutes."

"I am so glad that I met you," the first one said. Then he added after a pause, "We have half a million of our people living there amongst you. You wouldn't recognize them, however. They have been there for a long time, and we cannot desert them by allowing them to be slaughtered by idiots, so we take care of and look after them."

"Do they know they are Atlanteans?"

"No. Just a few do – those in the right positions and in the right governments, whose responsibility it is to make sure that no one makes a mistake of that kind. That is top secret; our people work directly with them. Your scientists are now much more educated and intelligent; they can understand our language and what we are trying to tell them. This is why they have such sophisticated technology, not to mention defensive weapons."

"For all of our sakes, for my sake and others like me, I hope you keep humans in line and make sure they behave as intellectuals and not as animals."

"Oh, that is without question."

After a brief word or two, they said that they were running late for a meeting and had to excuse themselves. They briefly said hello to the others in our group and then left.

Once they left, dessert was served, after which there was a repeat of the hand-washing ceremony. Once this was complete, Guyd appeared and said, "I'm here to escort you all to the art gallery."

36
A Visit to the Art Gallery and Library

All thirty of us tourists stepped out of the restaurant and took the short walk to the art gallery. I was in awe of not just the bronze-colored marble finish of the footpath we treaded, but the busts of celebrated humans that are perched on Greek columns at every crossroad. These busts are at every intersection. The busts are bronze, while the columns are olive in color, and are around three feet high. There is a bust positioned at a point on every side of the pedestrian crossing. In front of the bust is a plaque that depicts the individual's name and a small narrative on him.

I remember seeing some famous musicians, including Mozart and Tchaikovsky. It was obvious that this civilization proudly commemorates our achievements in the same way that parents commemorate the achievements of their kids, and proudly display their trophies and photos.

There was little traffic along the road; the only traffic I did see was in terms of people walking. Some shopping trolleys passed us by. The shopping trolleys travel at a speed of about forty kilometers an hour. Those in the shopping trolleys stared at us, always with a smile. I remember seeing a young girl, who looked my age, tease me as she passed by; she waved and then blew me a kiss. I flushed with embarrassment.

We turned right into a no-through street. This section of street has shops that are modern in flavor. I cannot explain how appealing the shop designs were to me.

At the end of the no-through street, they have a building that looks like a cathedral, and it has a sign with letters written in Old English, "Art Gallery." The architecture of the building is Roman, no different to what we would see in classical Rome. There are statues strategically placed outside the gallery. I saw Shakespeare on one side and a Greek philosopher on the other, whose name I cannot recall. These are full-body statues. One thing about the building is that it has no stairs at its entrance.

We all walked in and stood beside a magnificent Roman-style fountain, which has three tiers to it. The same two guides welcomed us.

It is impossible to picture the way the gallery is presented in terms of its overall layout, design, lighting system, presentation of the oil paintings, and décor. The ceilings are extraordinary. I saw domes with designs, and lights shining upon them. Some rooms have furniture in Louis IV style. In each room there is a table, about five feet long. In some rooms there are quilted Queen Anne furniture pieces, on which you can sit and admire the works of art. On the tables are book-form directories. These are in full color, and provide information, including an abridged history, on all the works of art, along with a biographical note on the artist. These are available to you if you wish to take one, in which case you are provided with a special leather case for it, which zips up. One man from our group, who was carrying a leather brief case, was collecting all the literature that he could.

At one stage we took a short break. I took the opportunity to approach our tour guide. I wanted to have a short word with her. "Excuse me, Ma'am. You'll forgive my ignorance, but I do have an ignorant question to ask you."

"I went through your data the first moment I saw you, and I could not detect any ignorance in your system. Nevertheless, what is bothering you?" I laughed, and she laughed back.

"The planet I come from . . ."

"I know which one."

"We have galleries, the same as you have, but nothing can match the expertise of your civilization. Many of the people I see in here are from my planet. There are lots from somewhere else, but to make a long story short, can you tell me if any of the thousands of artists who have created these works of art by the strokes of a brush on a canvas are still alive . . . or are they all dead?"

"It is difficult to refer to every print and explain who is alive and who is dead, but I would say that approximately half of the artists in this gallery are alive, except for the ones from your planet, who are all dead, some of whom have been dead for a long time. You know that our life span exceeds 1500 years, while your life span on planet Terra and many others in the universe is only up to around 100 years. Some, 150 years – it varies according to the planet. We just happened to have found a way to eradicate many of the factors that lead to death and premature death. For instance, we have found a way to get rid of viruses and bacteria that dwell in your body and have the ability to kill you. These are vicious killers that your immune system cannot sustain a defense against, and these killers lurk in the air that you breathe and in the food that you eat. You don't see them. You have already seen the blue fog in your unit; you have experienced a shower; you have seen the blue fog in other facilities, such as the toilet; and you've seen that our food comes in the form of a tablet, which is bacteria-free. We hardly keep any fresh food because we can produce our food in a different form. Our civilization has grown out of such practices, and we have no desire to eat anything that is vulnerable to disease or the like.

"I'll give you a small example. On your planet you have many wars, and many humans are still backward in their education, technology, and everything else; by now they should have been twice as advanced as they are. We have helped you develop technologies that can expand the life span of your people. We aid your scientists in finding answers to many of your problems, including finding antidotes. However, on planet Terra, as on many other planets, you have a vicious cycle of wars: millions of people are left to die and bacteria perpetuate. There are many areas where humans and animals are rotting and creating new forms of dangerous bacteria.

"Unlike you, we have found those defenses, and those defenses allow us to live in peace. We have found a way to protect ourselves. Our medical research has provided us with the answers. But one thing we don't do is give deadly bacteria a chance to exist. That is, we don't give viruses or bacteria on the land or in the water the opportunity to multiply as your race does. Much of this is because of the stupidity of wars; in wars, you leave millions of carcasses everywhere, with blood that serves to feed bacterial species, which can mutate into quick and efficient killing machines. In many cases, humans have unwittingly developed these themselves, and then they can't get rid of them. I would estimate that over half develop from this natural process.

"But we'll cut the conversation for the time being, and continue it next time when you come back; I'm sure that you will. Does that answer your question? I'm sorry that I don't have much time to explain everything to you as you may have wished, but I think you've got the message, and I don't think you'll forget it for as long as you live." She tapped me on the shoulder, walked a few feet forward, and then turned back and smiled at me. I stood there as though I were a statue and tried to comprehend everything she said. Now that I look back, her words have validity. We have had two world wars in which millions died. How much blood has been spilt, and how many parasites have been created as a result? I would say that diseases have killed more people than humans themselves have managed to kill, and that says something.

Some of those in my group were standing next to me, listening to our verbal conversation. What was amazing to me was that some did not understand anything. Nothing! During our conversation, there were occasions that two of them tried to interrupt us with a related question. My response was to kick the leg of anyone near me who interrupted us. To me, the conversation was something money could not buy.

The stupidity of some of these people was evident by one of their comments, "Where is planet Terra? I have never seen anyone like you. Your people must be very intelligent, the same as these people."

What was frightening was that he wasn't being sarcastic; he meant everything he said. I could only shake my head and say, "Well, on planet Terra, people like you belong in the zoo; from time to time people will throw a banana to you, just as they do to a monkey. Have you seen how quick a monkey is able to eat a banana and then ask for another?"

"You mean, that is how stupid we are?"

"If you have not been able to see this by now, you probably never will in this life, and probably not in your next life! Perhaps in your next life you will be born a python snake and live from the land."

"What is a python snake?"

"It is a snake that is long and all muscle. The first thing it's going to do when it gets you is wrap itself around you and squash every bone of yours. So that you don't feel much pain, it will sedate you. Slowly, you are going to end up in its stomach, and a special acid is going to dissolve you into some kind of liquid and other matter. After a couple of days, when the snake goes to the toilet around a tree somewhere, here comes Mr. . . . . oh, what is your name?"

"No, no, you don't want to know my name!" He walked off to the front of the group, took a handkerchief from his pocket, and wiped his mouth.

Having this conversation was fun to me. We must remember that I was a young man. This was not a telepathic conversation. I took this man and some of those with him as being primitive. The only thing I could assume was that the Atlanteans must have picked them up for experimental purposes. I didn't want to tell these people that, because I was worried that I may have regretted my words later. I thought that if these people found out, they might have taken offence and not only revolted, but become animals, as humans are noted for becoming. Even though they were not from Earth, but from another planet, I assumed that their primitive instincts would have been similar to ours.

One thing I noticed was that since I made the python comment, everyone in the group steered clear of me, which was a good thing as far as I was concerned. These people reminded me on the people I grew up around, which caused me to question myself in the following way: How in the hell do I have to always wind up in a civilization like the one we have on Earth? From young, I have always felt that I do not belong here. From the time I could think, I wanted to be up there in the stars; I have always known that my home is up there somewhere, and certainly not with the peasants I was born amongst.

We continued with the tour, and my mind was enthralled not just by what I was seeing, but by the knowledge that the tour guide was transferring directly into my brain, in much the same way as you would download data into a computer.

The next part of our tour was to visit the library, which is next to the art gallery. Upon our entrance into the foyer, a voice sounded from a speaker, "Ladies and gentlemen, on behalf of William Shakespeare and the dark lady, (name provided, to be revealed in another book), welcome to our library."

There before me, with their backs to me, were lifelike full-body statues of William Shakespeare and the dark lady. William Shakespeare is depicted with a mustache and beard, at around fifty years of age. He is wearing a top hat and is holding a walking stick. The tour guide said, "This was his favorite walking stick, and he used to always polish the brass." The walking stick is black, and it has a brass lion's head on top. Hanging from his pocket is a gold pocket-watch. His right hand is in his trouser pocket. His left arm is across his stomach, and he is holding his walking stick with his thumb on top. On this arm is the arm of a woman. She is standing at a right angle to him; her left foot is forward, just as his right foot is. Her face is turned to the left, in the direction in which Shakespeare is looking. She is dressed in a black cape with the hood down, and is leaning on him. Her left hand is on her right hand, similar to the Mona Lisa pose, while her right arm embraces Shakespeare's left arm. She has a slight smile on her face. He too has a slight smile.

The tour guide said, "In public, she always wore a dark cape with a hood." On this occasion her hair is in curls, which falls across her shoulders and down upon her chest. She has a touch of lipstick on, and a little makeup to lighten her skin. She is tall, even without the high heel shoes. Her stance is that of a stallion. Anyone who envisions her statue has no option but to shake. "In her time, people used to fall to the floor when they beheld her. Some shook as though they were having a heart attack. She had an amazing power. If she kissed any man that came close to her, he would have unhesitatingly given his kingdom to her, had he a kingdom to give. Her statue is a lifelike representation of her beauty." I agreed with the tour guide's statement about her beauty. No woman on our planet has come close to matching, and it is impossible that any woman ever will match, her beauty.

Many in the group agreed with the tour guide's comments and said that they had not seen such a beautiful woman on their planet. What a man would do for that woman!

Positioned about two feet off the ground, on a gold stand, is a six by eight inch plaque with a dedication to William Shakespeare and his companion.

Each of their feet is on a ten inch square gold plate, which is on a six foot square gold platform that can rotate. In the same foyer, there is a huge diamond chandelier. There is a small table next to the statues; on it are leaflets that provide additional information.

In this literary section, there are books on pedestals, and on the wall directly above the pedestals are oil paintings of the authors. There is a spotlight that gives off an amber and sunset glow of light. On the pedestals are plaques with dedications to the authors.

There are so many there, but Shakespeare stood out most for me. They have all of his works, even the ones lost to us because of censorship in his day. Of these we are not even aware; of his life we know even less.

It is difficult to absorb the way both the art gallery and the library have been put together, let alone write of it. To my mind, the architects and interior designers had to be geniuses. The art gallery is a monumental artwork in its own right. The main chandelier at the entrance of the library, and the statues of Shakespeare and the dark lady, are breathtaking.

The library is unusual in design; it is the future of libraries, as fantastic as it may appear to us. I saw rows of shelves filled with books. When you want to access a book that is out of reach, all you have to do is point your finger to the section with the book, and then make a downward motion with your finger. The bookshelf can read your finger gesture. Additionally, it does not need to rely on the finger gesture as the technology involved has telepathic capabilities. What will happen is that a section of the bookshelf, which is about one and a half feet in width, will lower to your height, and the lower shelves will disappear below the floor. The book you want will then move half way out so that you can easily take it. A written narrative on the history of the author of the book, and a picture of the author, appear as a holographic image.

You are able to pull up a chair and sit at the shelf from which you took the book. If you wish, you can read the book in the form of a holographic image. The technology that creates the holographic image has telepathic capabilities. This means that as you are reading the holographic book, the pages will turn at the appropriate time. Alternatively, you could sit there and read the actual book. As for the holographic image that appears, it is in the same size and format as the book. Indeed, it is a replica of it.

When you take the book away to read it, the shelf that lowered will return to its original state, just with an empty space in the shelf. When you want to return the book to the shelf, you point your finger once again and motion the shelf to lower. Before you will be the empty slot. After you place the book back into its slot, the shelf will return to its regular position. My immediate thoughts were that working in such a library would have to be a pleasure, and that there would never be a need of climbing a stepladder to reach a book again.

What is important for us to know, and an utter relief, is that not any of our ancient books that are of any literary worth have been lost: they are preserved in their library. We should note that only those books which the Atlanteans feel are of value are given a place in their library. To be in this library, a book has to have meaning and relevance.

37

A Historical Look at the Empire of Atlantis

My visit to the art gallery and library would not have been complete without a visit to the section dedicated to the history of the Atlanteans.

In terms of their works of art, they are far superior to our works. When you look at one of their oil paintings, you become a part of the picture, and you believe you are there. The portraits feel as though they are about to talk to you. Nothing on Earth closely resembles their works, apart from Leonardo's Mona Lisa. What makes Leonardo's Mona Lisa special is the energy he deliberately put in it. His energy was far superior to the energy of a normal human. This energy makes a painting alive, and it is part of the clue to deciphering the mystery of the Mona Lisa. The physics of this is explained in another book.

I saw statues of their inventors, and even of their politicians who contributed to the development of the Atlantean Empire. I observed that not even one of our politicians from Earth was considered worthy of representation in the same way, in terms of contributing to the development of our civilization. What a naïve question I asked the tour guide! "Why are our politicians from the surface not represented in a similar way?"

Her answer did not surprise me at all, "Because they are the biggest crooks and liars, and they are corrupt. They don't think of the small man who works for a living, whatsoever. If any of them were ever brought here, we would hang them within twenty-four hours, because our robots cannot tolerate liars, just as humans here cannot."

"I think you hit the nail on the head."

I couldn't believe it when she started patting her hands on her head. She looked at me and said, "I don't have a nail on my head."

I laughed and said, "You understood that literally. But, this is just a metaphor. Tells you much about our politicians!"

She got the message and said with a smile, "Oh, yes, I understand!"

We continued the tour. The tour guide said, "In this gallery we will be able to see, in pictures, the history of the Empire of Atlantis, as painted by our famous artists." Their history was just unbelievable. We were told of the process by which they travel throughout the universe, in search of suitable planets to leave not just their footprints, but human civilizations modeled on their own. She said that planet Terra was one such planet, and that they have been here for around twelve thousand years, in which time they have contributed to our intellectual and technological development, often by giving it to us on a platter. The same applies to other planets throughout the universe. On those planets, they also have colonies, unbeknown to the resident civilization. The way they guide our civilization is the method by which they guide these other civilizations. That is, they guide the evolution of the civilization by providing knowledge that humans cannot ever dream of. "On planet Terra, humans are just beginning to open their eyes – thanks to us. Other civilizations we are guiding are also opening their eyes. Some have developed at a quicker pace, and have long ago opened their eyes."

This concept alone made me feel as though my head were going to explode, like a balloon when poked with a pin. There was so much knowledge to absorb. I was resolute that next time I came to this place I would expand upon the knowledge I had thus far obtained. I was certain that I would return there one day.

The tour guide could only give us so much knowledge, knowing that we could not absorb everything in one go. She knew how much knowledge each of us could swallow, and this guided how much knowledge she transferred to us by the technique of knowledge transference. Transferring knowledge directly to a brain can overload it if the transferor is not careful. The first sign of overload is feeling light-headed, which progresses into a feeling of seasickness. The tour guide only gave us a brief overview. She told us about their planet of origin. She told us the names of all the planets they are colonizing. In the final analysis, primitive developing civilizations have a parent civilization secretly inhabiting their planet, and overseeing their development.

Once our two groups had regrouped at the entrance, one of the tour guides said, "Well then, I hope you are pleased with our small history, which you were given on this one-day tour." Everyone laughed at her assessment of the day. Everyone was saying that it was all mind-blowing knowledge. Everyone commented on how he wished to live in this paradise.

The other tour guide then added, "Well then, I understand that tomorrow some of you will be leaving us. We have decided to hold a special event in your honor tonight, in our auditorium, and you are all invited to attend. This will be a large banquet, with music, the best orchestra, and around a thousand people in attendance. This will be a unique and unforgettable experience. We already know what kind of menu each of you would most appreciate, so we will serve you this cuisine. Afterward, you'll open the dance floor. Each one of you will have a chance to perform your favorite waltz or dance to your favorite music. And you'll have the opportunity to choose a partner – one who most shines in your eyes! But we'll leave that one for tonight. Right now, from us tour guides, we say goodbye, and hope that you go in peace."

At this moment, Guyd appeared next to me and tapped me on the shoulder. I said, "Well, I didn't think I was going to see you again."

"You'll never get rid of me. I've been assigned to you, and I'll be with you until you leave."

Everyone in the tour group exchanged farewell greetings with one another and parted. Most of those in my group had their own guide, just as I had. I was tired both physically and mentally. Even though I had the body of a robot, it was programmed to behave in the way that a regular human body behaves; in this instance, my legs were sore from being on my feet and walking all day. The basis of how this works is psychological.

As we headed back on foot, Guyd said, "After the occasion tonight, you will go back to your quarters and prepare to go to sleep as you normally would. But when you wake up, you will be back in your physical body, and what you have experienced here will be just a dream to you, until next time we see each other." I was sad at this thought. I can tell you, it was hard to leave that place – particularly in view of what was in store for me when I returned to my near-dead body. Beyond my physical condition, I thought of the country from which I had escaped. Then I thought of how indefinite my future was. I had risked my life and fled from the only country I knew, and was going to have to find a new country to call home. Added to that, I knew that I was going to set out on this journey on my own. For a young man this was both terrifying and exciting: exciting in that the life I was going to write could only be better than the life I had left behind. I had so much enthusiasm, and I was so intelligent, but fate has an interesting way of changing your plans and writing a path of its own. Whatever that path would be, I knew that one day I was going to tell humans that we are not alone. That there is more out there. That the deception that suppresses our ability to be real humans and live as humans should, has to end. That those who enjoy the fruits of this deception are kidding themselves, because their actions are noted. I am pleased to say that fate has taken me on a road that is enabling me to do this.

38

A Farewell Celebration

Greeting me in my quarters were four beautiful girls, who looked no older than nineteen. They had blond hair and pigtails. At the time, I was wondering whether they were robots or humans, as I couldn't tell. But my instincts told me they were robots. They said that they were going to help prepare me for tonight's event. One of the girls said that she would groom me, and another girl said that they would leave me in her hands until she was finished. I went to the bathroom, where there was a barber's chair in front of the vanity mirror. She said, "I am your barber, and I will shave you, cut your hair, and prepare you for your grand finale, which is in two hours."

Once she completed grooming me, she said, "I wish you luck and happiness tonight."

One of the other girls returned into my unit and said, "I am your tailor. I have brought half a dozen different outfits for you to choose from, ranging from the 18th to the 20th centuries . . . Oh, if you let me choose, you cannot go wrong." I agreed. She chose an outfit from the 18th century. She added, "Oh, don't forget to have a shower before you dress."

"Thanks for reminding me. I would have forgotten."

"I will put these away in the wardrobe, and then see you in half an hour or so."

Now in the bathroom, I undressed and tossed my clothes in the dirty-laundry basket. Then I went into the shower, which only lasted a minute and a half. I remember hearing the noise of running water in my brain. I felt as though water were falling all over me; but when I opened my eyes I could see a faint blue mist all around me. I just stood there. Sometimes I touched myself to see if I was wet; in each instance I discovered that I wasn't.

The blue fog disappeared, and a voice in my mind said, "Your shower treatment is complete."

I stepped out of the shower and walked back into the bedroom, nude. On the bed was my outfit, which I dressed into. I always found it fascinating the way clothes altered themselves to fit my body. Then I went up to the mirror to see how I looked, and I liked my outfit. Unfortunately, my hairdo didn't survive the shower. When I turned back I saw a third girl in the room. She said, "Well now, let's sit down. I'm the makeup artist, and I'll fix your hair, trim the hair in your ears and nose, and give you a touch of makeup if you wish."

In no time at all my hair was done; she even curled it a little to go with the period costume. I looked like Napoleon. Everything about the way I dressed and looked was done to perfection. Even my nails were done, and perfume was put in the right places.

Once I was ready to go, Guyd appeared and surprised me. He, too, was wearing a costume from the 18th century. "I shall escort you there; once there, someone will escort you to your seat."

We took the lift to the ground floor, and then walked for about five minutes, until we reached a large building that was well lit. At its entrance were two doormen dressed in 18th century French costumes. A third doorman stood in the middle of the doorway. He pounded his staff several times on the floor. The staff was made of redwood, and there was a diamond on top. With this action, the orchestra momentarily paused. The music was not audible but telepathic. An attendant appeared before me and bowed with a slight nod. I bowed back. He said, "This way." I followed him to the best seat in the house. Everyone in the auditorium stood and clapped in my honor as I walked to my seat. Those at my table smiled as I sat with them. On my left was a young couple; on my right, an elderly couple. The auditorium was practically full.

I noticed that everyone wore period costumes and jewelry. Until that moment, I had not seen an item of jewelry worn by anyone.

I looked around, and could hear all the conversations going on. I even heard people speaking in different languages. German was one language that I recognized. A number of them spoke in English. I discovered that when one of these people meets you, he penetrates your mind to ascertain the language you speak, and then he communicates with you in that language.

It was a happy atmosphere. People were all over the place, conversing with one another; the music of Strauss was playing. A gentleman stood on a podium and made an announcement. He did not use a microphone. With telepathy, you have no need of a microphone. He said, "Ladies and gentlemen, I am pleased to introduce our guests tonight. This will be their last evening with us, so we have decided to hold this banquet in their honor. We have from planet Terra, this gentleman, Mr. Jack Lord . . ." He pointed me out and the audience applauded. Then he mentioned several others, who were scattered around the room.

A ten-course dinner ensued. Each course was portioned so that you had room for the remaining courses. Guyd knew me well. From across the room, he said, "Don't eat too much because there are a lot of other dishes on the way; you won't have room for them all if you eat too much." Despite his warning, I found room for everything. Anyone watching me eat would say that this man has never seen food before. Only a man who has grown up in poverty would understand my eating habits. For dinner, I was served my favorite dish: Lobster Mornay. For dessert, I was served one of my favorite desserts: Black Forest cake.

The conversations I had during dinner involved planet Terra. The first thing someone asked me was, "How are our cousins doing there?"

"I don't know. I've never met one yet."

"One day you will." Everyone at the table laughed.

Liquors were served, and I helped myself to a green colored one. I had no idea what it was I was drinking.

After a while, the master of ceremonies stopped the music and said to the orchestra, "I'm sorry to interrupt you, but it is time to start our program. First, we have chosen a beautiful girl in our midst to open the dance floor. Her partner will be Mr. Jack Lord from planet Terra."

I heard the voice of Guyd say, "You go meet her in the center of the dance floor." I stood and slowly walked onto the dance floor. She did the same. We came from different sides of the room and met in the middle. She curtsied and I bowed. The music commenced. We put our arms around each other and waltzed to the tune of Strauss. The lights were dimmed once the music started. On this occasion the music was audible and not telepathic. I was so nervous that I was shaking, but then I said to myself, I'm supposed to be an actor! Why am I shaking – it's not the first time I've been in the spotlight? From that moment on a confidence overcame me, and I took command of the dance floor – after all, I was a dancer, and a good one at that!

The dance floor had a surface that looked like glass, and from it emanated dimly lit colors that almost had a hypnotizing effect on me.

My dance partner was a blue-eyed blond, with wavy shoulder-length hair. Her hair was so shiny, but its shine did not match the gleam on her face when she smiled. She had a most beautiful dress on. Each time I swung her, her long dress fanned out like an open umbrella. You can only imagine her figure; her firm body looked as though it was made to order. But it was to her exposed cleavage that my eyes were drawn. Perhaps one could have been fooled into thinking that it was the magnificent black diamond which had me captivated. I felt as though her breasts were speaking to me, and I understood their language. I was naturally excited, and I know that she liked it that way. I felt that she was so happy in my arms.

We spoke to each other by voice, and I noticed that she spoke good English. I said, "Because you speak telepathically, you don't need a mouth. Maybe one day, thousands of years from now, you will lose your mouth because you're not using it."

She laughed, "No! We need a mouth all right. We use both systems. We have to eat with our mouth, and need our ears to hear, so we do need everything. We are just like you."

I must admit, I was only thinking about how I could get closer to her, in bed! At the time, I was so caught up in my emotions that I completely forgot that she could read my mind. I remembered afterward, but that was too late.

Each time I thought of something that I shouldn't have been thinking of, she would press her body close to mine. When we danced the tango, I cannot tell you what I was thinking! It didn't occur to me that she knew what I was thinking. Then again, just as she could read my mind could I read her mind. My impression was that her thoughts were synchronous with my thoughts. I was being taken over by my fantasies, and when I looked into her face, all I could see was how red she had turned. The whole experience was unusual, to say the least. Added to that, she was very cheeky and knew how to make you laugh. She was full of life and had a vibrant personality. Had it been my destiny to stay there, I probably would have married that girl, and she would have definitely married me, had I asked her.

I hadn't even noticed that others had joined us on the dance floor. One must wonder why everyone in the room was not in uproar, considering what my intimate thoughts were. Thank goodness couples have a telepathic privacy cut-off feature in their brain. When couples have intimate thoughts, their telepathic frequencies are automatically isolated. This means that their shared intimate thoughts remain private. I would hate to think what people would have thought had they known what our thoughts – at least what my thoughts – were while we were dancing. Some things in life are best kept private.

It is interesting, however, because not all sexual thoughts are kept private. Let us take a hypothetical situation. If you desire someone else's wife, then everyone will know your thoughts – including the husband of the wife you desire. This hypothetical scenario has identified the limitations of their privacy. The one narrow margin of privacy involves appropriate intimacy. Appropriate is an important word here.

You may say that the two of us were having intimate relations with our minds. I felt that she was participating in my fantasies, by making them her fantasies. I also felt that she was adding her own fantasy to my fantasy. Perhaps I was mistaken. It is hard to explain this, but I felt that your thoughts become one in this instance. I considered this a sexual encounter of the fourth kind! At one point her thoughts were, "I wish I could find a man like him. I'd be the happiest person." I pretended that I didn't hear her thoughts, and was deliberately looking left and right, acting as though I was not paying attention to them.

After spending around twenty minutes on the dance floor, the music stopped, which prompted me to return her to her seat. I pulled out her chair, and after she sat I kissed her hand, bowed, and said goodbye. When I looked up at her, I saw shiny eyes and a red face. I knew everything. I said, "Perhaps we'll see each other again before I leave."

"I'll try hard to see you again," she solemnly said, with a defeatist attitude. I walked away and didn't speak to her again. We just kept exchanging glances all night. She sat with her back to me, and often turned around and glanced at me. Her friends did the same and giggled. On one occasion when we exchanged glances, I picked up the thoughts of one of her girlfriends. She said to her, "Ask him if he has any brothers." I laughed and thought what a shock she would have knowing what kind of family I had. I truly wanted to stay there, and was sorry knowing that I had to leave.

Just as she had to dance with other partners for the remainder of the night, so did I, but I wanted to be with her. Unfortunately, the opportunity did not present itself. I had to make myself available to all of the other women; to dance with her all night would not have been diplomatic let alone appropriate.

Women wanted to dance with me, while men wanted to talk to me. Thus, when I was not dancing I was having conversations about planet Terra. Even though Atlantis is on our planet, when you are in their world, planet Terra seems to be light years away, almost in another universe.

I knew that I was making a big impression with the guests. I spoke of what planet Terra is like, and many people told me that they would definitely come to visit planet Terra. One said, "I will make a point of visiting your big cities and shopping centers."

"When you do, will you please call on me?"

"Yes, no problem."

I learned that they have not only day excursions to the surface, but overnight ones. They even have holidays here. There are always protocols involved, and there are special robots that accompany the guests in all of these instances. You cannot go to the surface on your own. If this group happened to be in Sydney, climbing the harbor-bridge, and you saw them, you would never know. They come prepared, with money and with documents. You may have already met someone from Atlantis, and you will not know that you had.

On their part, first contact was made a long time ago. On our part, I made first contact on that visit to the mother ship.

The tables were full of cakes, drinks, and food, right to the end. The attendants didn't begin to clear the tables until the last person left. The lights were turned up as the clock chimed midnight. This was a grand old grandfather clock. I exchanged countless greetings on the way out. Many smiled at me and wished me well. All expressed the desire that I come back again.

I was a little tipsy when Guyd and I strolled back to my unit, which I thought was a little odd since alcohol is not supposed to have an intoxicating effect on the brain. Apparently, being tipsy is psychological. I was even moving a little to the left and a little to the right.

Guyd teased me about the girl on the way back. "Did you try her figure out by putting your hand down a little from behind? Was it hard, or was it soft like a marshmallow?"

"Well, that's something I will carry to my grave and not tell anyone. You have a very nosy mind, don't you?"

He laughed. "Well, at least I know one thing: she's in love with you. I telepathically heard her telling her girlfriend at her table. 'Love at first sight.' That's what someone said."

"Well, it looks as if she gave me the same virus, because I feel just the same. Maybe it's because I'm drunk from two glasses of liqueur."

"No! Whatever happened was real."

Once we arrived back at my unit, Guyd said, "You know what to do to get your bed out and the other stuff?"

"Yes, I just say, 'Bed out, and so on.'"

I was thinking of how I needed to have a shower because I stunk, and he said, "Well, I'll leave you to your shower. I probably won't ever see you again, but you've been fun to be with."

"I'll come back again one day."

"Oh! Then I'll be waiting for you."

"What for?"

"I've got to know if her backside was hard or soft like a marshmallow!"

I grabbed him by the shoulders and turned him around so that he was facing the lift. "Now you go get lost." I pushed him into the lift.

We hugged, and he said, "Until next time. And I know you'll be back. Your body will be kept here for you, until your return."

"I've made some friends tonight, who said they will probably come to see me."

"They probably will." He hesitated and then said, "Good night." We shook hands, and I watched the lift disappear. I knew that Guyd was genuinely sad to see me leave. In my short stay, we had touched each other's hearts and become true friends.

39

My Body Wakes from a Coma

I went into the bathroom, undressed, tossed all those beautiful clothes into the dirty-laundry basket, and took a shower.

The temperature in the bedroom was comfortable. The bed and pillow were so comfortable to lie on. My feelings were beyond words. How does one explain what it feels like to live in this place? How does one even describe a feeling? I didn't want to close my eyes because I just didn't want to wake up in a different bed. I fought hard to keep my eyes open, but sleep was fighting a battle with me that I knew I could not win. Sleep had more to offer than I wanted at that moment. But I was so exhausted that no matter how much I fought to stay awake, I couldn't. I fell asleep; when I awoke, it was in a different bed, in a different place. The sad thing was that it was with no recollection of anything that happened. Such a beautiful encounter was lost to me for most of my life. Only in the year of this book being written did it all come back to me.

BOOK II

A PERMANENT COLONY IN THE PACIFIC OCEAN

Introduction to Book II

Before I proceed with the details of this visit, I should make it clear that, although I had visited the mother ship in the Atlantic Ocean while I was in a coma, I had no recollection of this visit once I returned to my body. Just as past lives are vaulted in the subconscious mind, lost to the conscious mind, so this knowledge was stored away in my subconscious mind and was lost to me for most of my life. Timing is interesting; fate is instrumental in this, for, fate is often a factor in the release of knowledge from one's brain. To some, the release of knowledge can determine the course of their lives. It is always written when your brain will release knowledge to you, and what that knowledge will be; sometimes that knowledge is life altering, not just for the bearer of the knowledge, but for society as a whole. When knowledge has been designated for release, something automatically "switches on" in your brain. It is as though there is a program in your brain written by fate set to activate at a key moment in your life.

When I returned to my body after the second visit, I recollected only part of the visit. When did fate open the door for me to access the comprehensive details? In 2011, when Rosemary began writing *The First Cause, Volume II.* The account of my visit to Atlantis was only meant to be a small chapter that contains my astral travel experiences. She read to me the recount I gave the day after the 1994 experience, which she had recorded on cassette tape. This was when a command in my brain set my subconscious mind to access and release the knowledge of my first visit to Atlantis. The more I remembered, the more this chapter became a book in its own right. So was the inception of this book.

As we know, our subconscious knowledge is barely accessible to us in our present state as humans. You can look upon your brain as being a computer, with lots of files, with those files representing different lives and different categories of knowledge. In my case, I can add to that files of different astral travel encounters. While I have access to my subconscious knowledge, I have to search for the right file. However, first I have to know that the file exists! That is the important factor: knowing the file exists to access the knowledge. You can compare this to a dream. If you have a clue to the dream, you can build on that clue and try to "pull" the rest of the dream from your subconscious mind. Sometimes a word or a thought can trigger an association with a dream, which is your clue to retrieving it. Why you cannot remember it is that it has been filed away somewhere in your subconscious mind. With a clue, you may be able to draw it out; however, without a clue to what was in the dream, the likelihood of remembering it does not exist. Then, what is the likelihood of remembering a dream decades later, when you could not even remember it the day after? You have to know what file it is you want to download. This is what happened to me in 2011. My brain (there is more to this, and this is discussed in the biography of my life) made me aware of a file in my subconscious mind, and then I was able to access the knowledge in that file. The trigger, or command, for me to become aware of the file was Rosemary writing the second volume of her book. Fate had it written that I release the knowledge to her at that time and not a moment before. Once I initially opened the file, I was able to draw out as much as I physically could, considering my state of health.

In this life, in my astral state, I have made many visits to other worlds. It is only for my good fortune of telling Rosemary, who at some stage found the wisdom to record my recounts. I cannot remember many of these, until I am reminded of them. In some instances, I still cannot recollect them. Once knowledge goes into the vast recesses of the subconscious mind, it is difficult for a human mind in our present state of existence to draw that knowledge out. In my case, from birth I was gifted with unusual savant-like skills that have developed to the extent that now, in my twilight years, I can have access to my subconscious knowledge; what is absurd is that there is one requirement: energy. Why this is absurd is that by the time you have evolved your brain, you are usually in your twilight years, and it is in these years that your energy is wasted on health problems, which means your energy is available for little else. By diverting energy from where it is needed, you pay a price in terms of your health. The brake on my potential is my health. What I could do with my brain if I had a healthy body is not worth speculating on, because that is what is wrong with our world. By the time you evolve your brain, your body is the brake that prevents you from utilizing it as you otherwise could.

The governing authority of Atlantis knew that through me the first stage of contact would occur with humans on this planet, just as through Plato knowledge of the existence of Atlantis was carried through history. In the *Conclusion,* there is a brief reference to how the second stage of contact will occur.

1
My Astral Travel Visit to the Colony in the Pacific Ocean

The year was 1994. My visit to the colony was prearranged. We would not understand the details of how it was prearranged at this point without background knowledge, which can be found in another book, so this aspect is best left out of this work. What was prearranged was that one of their people would allow me to leap into his brain for the short duration of my visit there. When you leave your physical body by means of astral travel and plan to leap into someone else's body, you have to have a body available to accept you. There are many factors at play here. (The process of leaping is explained in another book.) Once you are able to take possession of a body by this means, you then have the ability to communicate with the physical world. When you leap out of that body, your memories and knowledge obtained during the visit are taken with you.

Where was I going? To a colony of space people in the Pacific Ocean. This is the unknown part of the legendary Atlantis.

Out of my body, at the speed of thought I found myself in unfamiliar surroundings. I knew where I was, as the knowledge was transferred into my mind. I had leaped into somebody's body from this colony. With this body I had the full extent of telepathic capabilities. I was in a room, sitting on a lounge, and there were two other men present in the room. The lounges were modern. There were some paintings here and there, and television screens. It may seem strange, but we did not exchange greetings as you may expect. This is because these formalities were unnecessary, as we all knew who was who. They were dressed in trench coats and hats. The trench coats were beige in color. Their style of attire was old-fashioned. They appeared to me as being private detectives from the forties era. I was dressed in the same attire.

These two men were governors of the colony, and the body I temporarily possessed belonged to a governor.

We never spoke audibly. We thought; telepathic knowledge was exchanged by way of direct telepathic conversations, and by way of knowledge transference. My questions and curiosities were answered in my mind as soon as I posed them in my mind. When I wanted to ask a question, I looked at them and they looked at me, and I knew the answers in my mind – this was without even having asked the question.

One of the men said, "Let's take a small tour, and I'll show you something unimaginable that you have never seen before."

We walked through a corridor and stopped in front of a door. One of the men knocked on the door before us, but we did not walk in. He told me that they have the ability to walk through walls, through doors, or through any matter, but they don't practice that out of common courtesy and respect for the privacy of others. This ability to walk through matter is only applicable to the governing authority and not to the humanoid robots, which do not have a soul, let alone to the general population. I watched as one of them demonstrated this to me. I saw his physical structure go from a matter structure to a non-matter structure – an energy state. In that second, his image just appeared pixilated, or fog-like. It was in that state that he passed through the wall. Moments later, he opened the door from inside the room and asked us to come in. He said that when he had crossed through the wall to the other side, his body reversed the process in a matter of a second, in that it went from a fog-like state back to its regular physical state.

With their minds they can do anything – they can appear, disappear, walk through matter, and so on. (As stated earlier, this is only applicable to the governing authority.) In their position of authority, it is essential to possess such abilities.

The physics of how he did this may seem vague at this point, until we remember the energy state described in an earlier chapter. This is the energy state that can transcend matter. We learned of how you can request that your dinner be ready for you on the dining table. In that case, a tablet would transfer from the fridge to the food converter, from which your dinner would transfer to the dining table. In the process, the dinner would travel through the air in a fog-like state. One could conclude that this is the energy state that he was able to convert his body into. This book has not been able to define this energy state in terms of its subatomic particle nature at this present time. The soul is an example of a different energy state that interacts with the matter state. (Another book has defined the soul in terms of its subatomic particle nature.)

The door opened to an office. There were several lounge chairs and a large desk. Behind the desk were three girls dressed in uniform. They wore a white pantsuit, which had thick (two inches thick) blue stripes. Zippers went up the sleeves to a skivvy-like collar. The pantsuits were body fitting, revealing slim builds. With the positions they held, I knew they had to be robots. Everything that I saw in this office was modern in style. While the look was modern, the furniture and the frames of the pictures on the walls were unusual and not like anything I had ever seen. The framed prints were of a cosmological nature.

2
A Big Brother Society

The classical music that I heard in my mind made me feel serene and at peace. When you develop telepathic capabilities, your mind becomes the equivalent of a radio receiver, which means that your mind can tune in on any frequency of music that is available for you to tune in on. Whereas we have radios that pick up the frequencies of radio transmissions, they have their minds. It is interesting because the music I was tuned in on did not interfere with the conversation I was having. It was soothing in the background of my mind. Just as everyone can tune in on whatever music he wants to hear, can everyone tune off it.

I saw a large-screen television, and at least two types of recorders on shelves, which were emitting flashing lights. I was curious, so I asked one of the women what they were. She said, "These are tuned in on peoples' conversations, no matter how many miles away – including the one we are presently having. We tune in on broadcasts from other planets as well as from your planet."

This television screen was holographic, and it appeared and disappeared at your command. This was demonstrated to me. One of the two men clicked his fingers, which caused the screen to disappear, and then with another click it reappeared.

He showed me a sample of what kind of programs their people watch. He said, "Everyone loves *The Three Stooges.* This show has never gone out of fashion here. While we find the characters to be very stupid, we find the show very funny to watch.

We watch many of your television shows. Of particular interest to us are comedy shows. You will not believe it, but when the Beatles came on the scene, we fell in love with their performances – the way they jumped around. They made us laugh so much. The Beatles reminded us on *The Three Stooges*. Their music has not gone out of fashion but still appeals to us, as does watching clips of their performances. Nothing, though, beats seeing monkeys peel bananas. There is something in the way a monkey uses his intelligence to peel a banana and then eat it that we find fascinating. We even have a common saying that is bandied around in the workforce: You work just like a monkey peels a banana."

I discovered that they also watch reality-television shows, but their programs are not the same as our programs. The stars of their reality-television shows do not know that they are being recorded or watched. Therefore, the "actors" do not behave to the camera. This type of program involves ordinary humans living on the surface – namely, us, in our homes, how we eat, how we sleep, how we behave, and so on. In other words, they are watching us in their living rooms. (Earlier, we learned that some of us are tagged. Those of us who are tagged fit into the category of being studied and watched.) Big brother from Atlantis is in our homes, in our work, in our streets, even in our toilets. There is nowhere for us to run or hide. Because we don't see them, or even know that they are watching us, we feel comfort in our privacy. Yet there is no such thing as privacy.

All they have to do is tune in on you. How do they do this? They do not need video cameras, although nanorobots and parts are their video cameras in many instances. All they have to do is think about you, and they are able to recreate your thoughts on the television screen so that whatever you see they see. Their technology can telepathically tune in on your mind, on your thoughts. They can then record everything through your eyes. The governing authority controls this, and determines who stars in the reality programs and what the viewers watch.

Frankly, we may be surprised to know that there are many things the citizens of their society are not permitted to see. Whatever they watch is censored. For instance, they censor extreme violence and gruesome sadistic acts, such as killers cutting heads off. They also censor immorally unacceptable behavior, such as sexual deviations and Sodom and Gomorrah style behavior. There are sexual behaviors that we see and in some cases permit in our society that are not permitted in theirs, and they cannot be viewed in any way. This means that pornography is banned.

Besides, these people would be embarrassed by such shows, and would never consider watching them. In their society, there is no immoral behavior; there is no such thing as prostitution; and there are no sexual perversions or deviations. This all comes down to the fact that the negative side has little if any input at all in the people of their society.

Where we presently were was the command center of the colony. From the command center, every person in the colony is monitored. This means that if someone is planning to sabotage the colony in some way, then his thoughts and actions are recorded. His every move is captured and can be viewed on screen.

What this tells us is that there is no such thing as a perfect society, no matter how evolved it is. These humans are, after all, from the same genetic seed as us, and they have the same negative and positive energies within them that we have. (This is explained in another book already mentioned.) This means that while they do have the rare troublemaker, it is impossible for anyone to sabotage the society or do harm to others. No one is given the opportunity to do serious damage because he is caught in the thought process.

What then happens to the rare troublemaker is that his mind is altered so that he does not tread down the same negative path again. Having negative thoughts means you are allowing the negative influence within you to steer you. Someone who has extreme use of his negative influence is considered to have a defect in his brain. Based on this, they must consider that many of us are defective. In our society, we have temptations. In their society, they remove temptations that offer opportunities of harming others or self-harm, wittingly or unwittingly. These people cannot even get drunk, smoke, or take drugs, because the temptation does not exist. There are no drunks, smokers, or drug addicts in their society. There is alcohol, but it does not have any of its mind-altering properties. People are kept on the straight path.

In our society, there are those who take advantage of others and offer temptations for self-serving benefits. This includes gambling. These things must one day be removed from society. Negative temptations must be removed from people who cannot help themselves. Moreover, this caliber of intelligence would not be interested in such things as poker machines or casinos. Having said this, if the temptation were there, there is always the possibility that someone could fall into the addiction trap, simply because the very ingredients within a human that drive him down the path of addiction exist within these people.

In a superior civilization, humans do not even know what it is to lie. There is no opportunity for them to lie. For one thing, the ability to read minds negates the possibility of telling a lie!

The lack of privacy in their society may seem daunting to those of us who value our privacy; the truth is that if you have nothing to hide, you have nothing to worry about. In our society, with our degree of intellectual development, a great many humans have much to be concerned about. The concern someone has of knowing that others can read his mind is an indication of his intelligence – with intelligence directly proportional to his use of the positive and negative influence that exists within him.

Returning to their reality-television shows, we should not feel uncomfortable; we should consider it in this way: just as we watch documentaries on the natural world, so others out there beyond our little cocoon watch documentaries on the human society here on Earth. Just as we may study the behavioral habits of animals, so others study the behavioral habits of us human "animals" on Earth. Superior races of humans view us in exactly the same way that we as humans view inferior species of life on this planet. We may or may not take pride in knowing that we provide a great deal of entertainment and interest to others out there, who are watching our evolution at every turn. This book is not just speaking of the civilization of Atlantis, but of any intelligent life in the universe. Besides, the documentaries, or reality-based programs, are transmitted not just to the homes of Atlanteans on this planet, but back to all of the planets they colonize, and to all of their neighboring planets. Frankly, they are transmitted to anyone who is able to pick up their signals. This means that you might be the biggest star out there in the universe, and you are not even being paid for it! Whether that would be flattering to you is another question.

There is no comparison between their so-called big brother apparatus and our so-called big brother apparatus, which we either have or can have. This is because humans are always capable of being tempted to the negative side, as the negative side is a part of a human's genetic makeup. This is our biggest fear and threat. In a society that has a governing instrumentality such as that of Atlantis, the citizens do not have any possibility of corruption in the upper echelons. First, this is because the metaphysical being in the robot of a governing authority comes from the first source of intelligence to have come to exist in the universe, and it has no negative qualities with which we are corrupted. Second, in his physical form of existence as a member of the governing authority, his physical body is the most sophisticated technology there is. This is why people don't think twice about loss of privacy.

In their society, monitoring of citizens ensures the safety of citizens. Let us consider an example. If you say something that is detrimental to the society, you will instantly appear on everyone's television screens – not just the screen in the command center. This means that the entire population witnesses a repeat of what you said and how you said it. Your intentions are exposed for all to know. Then the appropriate robots will pick you up and take you away for a special brainwashing session to remove the negativity from your brain. This way you will cease to have negative thoughts. They can wipe this specific negative energy from you. From then on you will not remember the negative episode, but become a good citizen, just like everyone else.

In their society, the negative side that we all have within us does not prevail or have the opportunity to prevail as it does in our society. Having only one or two of these negatively-inspired people in a population of one or two million is insignificant. In their society, for the most part, no one can do harm to anyone; there are no such things as liars, crooks, or cheats; there are no such things as sexual deviations, including child molestation; there simply are no seedy elements of society. Furthermore, there is no injustice and no corruption. This means that the people can trust those they need to trust. What this shows us is that the Atlanteans have as close to the perfect society as you can have.

3
A Description of the Colony

I sat on a lounge. I was offered coffee, tea, a hot drink, or anything I wanted. When I asked for a hot chocolate, it automatically appeared on the table before me. I asked for a hot chocolate because I remembered how much I loved it as a kid. The most important thing for me was that it had filled me up and had helped me to forget about my hunger. This was a sentimental drink. I drank alone.

"What I really want to know is why I was brought here," I said.

"It is time you know how the destiny of our people fulfills the prophecy you made, that one quarter of the ocean would one day be occupied by us. Now you can find the answer to the question you posed at the time regarding how we would return here and where we would live.

"As you know, the Creators have always maintained balance and equilibrium in everything that exists, whatever that might be – the natural world of plants and animals, and even man. Order must exist for living beings to survive. So this should answer the question you also posed of why planet Terra is not balanced and yet exists. You were so right; until now, three-quarters water and one-quarter land has claimed planet Terra. As you predicted, one day we shall return and claim one-quarter of your ocean, where we shall reside. Only then will the Earth become balanced – in other words, half of the Earth shall possess land with land-bearing creatures, and half shall contain water with water-bearing creatures.

"You stated that we shall occupy one-quarter of the ocean, but we won't have to evaporate it. As you can see," he went on with a smile, "we don't have to go to the trouble of drying up the ocean and altering the planet's orbit. Instead, we can still live in the ocean, but, as you can see, in a totally different type of environment that has been created for our civilization, which corresponds to the environment on the surface of your planet."

There were six three-D television screens on the walls. Every screen depicted different images of the colony. One screen illustrated the architectural blueprint of the colony. Another screen depicted the colony as it presently looks. The colony is on the floor of the Pacific Ocean. Its size then was around half the size of Sydney. One of the men said, "Because the floor of the Pacific Ocean has vast regions of flat plains, it is an ideal place for a colony. Naturally, there were occasions that we had to carve out the ocean floor to make it flat."

I was viewing an aerial picture of it, of how it must appear when it is not camouflaged with marine life. I saw hundreds of roofs, all a brown rusty color, and all joined together. The roofs are hexagonal (six sided) in shape, and every single roof represents an area the size of a typical suburb. I noted that the roof design is an intelligent one, based on one of the strongest and most efficient shapes in nature, the hexagon, which is why bees use this shape to construct honeycombs in their beehive.

"There was no shortage of technology or material for us to complete the project. As the colony expands, another section is just added to the existing structure.

"The material that the roof and walls are made from is so tough that it can withstand extreme temperatures, akin to millions of degrees of heat, which includes volcanic activity; it is bullet proof and laser proof. Furthermore, it can withstand strong under-ocean currents and earthquakes. The material never rusts, and it never breaks. While this material cannot be found on this planet, it can be found on the moon. It is the hardest substance known to exist."

I estimated that the thickness of this material is approximately fifty meters. The same man added, "The mother ship is made of the same material, which enables it to fly into volcanic lava and heat of great magnitudes, with no consequences to the mother ship let alone to its occupants. The mother ship will never be affected by heat, or be scratched by meteors or even by larger objects, including a direct hit by a planetary object. In this instance, the mother ship will penetrate through that planetary object. You cannot just travel in the universe at tremendous speeds and not be able to protect yourself. You have too many people at risk. What you have is a traveling city in the mother ship.

"The same applies to the flying saucers that are attached to the mother ship. These flying saucers, by the way, can travel at one and a half times the speed of light. Any structure made of this material, no matter what its size, has the same strength and potential."

Each section of roof has a slight gradient that rises to a central apex. This apex is slightly rounded. From each roof there are pillars that extend deep into the ocean floor. The pillars are made of the same hard material used to make the walls and roof. As pillars support the roofs, you can see these pillars throughout the colony. These pillars are round in shape and large.

The dirt that was removed when the pillars were dug into the ocean floor was strategically placed on the rooftops. Marine life has turned the surface into a natural seascape. Radar or the like sent by us would not be able to detect their presence because the structure absorbs the signals. We simply do not have the technology to detect their presence in the ocean, until they want us to detect them.

There are no major structural walls inside the hexagonal structures, which we know are the size of an average suburb. The only walls that exist are those that separate the inside structure from the ocean – that is, the outer walls and the roof. Whenever the colony grows and a new section is added to the existing structure, the walls of the old section, which used to be exposed to the ocean, are removed. These are no longer necessary as the new section separates the ocean from the colony.

4
Fishing Areas

One screen had a scene that baffled me. I saw an area of the ocean floor with many dead fish, and hundreds of lobsters. People were walking around collecting them; they were putting them into what looked like rubbish bags. One of the two men said, "These are robots, and they have to be careful because, these lobsters, unlike fish, can live out of the ocean, so they are still alive."

My first instinct was that this region was some kind of fishing area. I asked, "Why are the fish dead? Where do they come from?" His answer provided particulars that I have not attempted to describe. The following is a basic account of the details, as I best understand them today.

"These regions are fishing grounds. We have created quite a few of these areas to collect fish and lobsters. We have converted this area of ocean into one that has the same atmospheric conditions that you have on the surface. We have also created a gravitational force in this region that coincides with the gravitational force you enjoy on the surface of your planet. We have pressurized the water with atmosphere. This is why you can walk in that area and breathe in the same amount of air as you can on the surface. We have been able to convert regions of the ocean into atmospheric-bearing regions."

"What would happen if a submarine or ship passes through this area? Would it just crash down into it as if it were falling from the top of a mountain?"

"These areas have no true barriers to keep fish or water out. They have ocean all around them and on top of them; however, seawater doesn't splash in on its own, but fish do. If a submarine passed through this region it could crash in, just as mountain rocks or volcanic lava could if we didn't screen them out. We allow things in that we want to come in. Such includes fish and lobsters, which we eat a lot of.

"So in answer to your question, yes, that's exactly what could happen. It will be a similar situation to the Bermuda Triangle. To prevent this, when we see a ship approaching the area, we create an electromagnetic field, which is a bridge for it to pass over the area. Because of this electromagnetic field, there could be thunderstorm activity in the area. The electromagnetic field acts as a roof. If we failed to produce the necessary electromagnetic field, submarines and the like passing through the area could be sucked into this region and fall to the floor, just as you said, like from a mountaintop.

"While these farming regions have not claimed any victims, the same cannot be said of the Bermuda Triangle. Having said this, in the later stages, we managed to stabilize that region so that we can avoid such things happening again – that is, lives, not to mention ships and planes, being lost."

I understood that these fishing areas are "pockets of air" on the bottom of the ocean. Just having an air pocket is not going to provide humans with the right conditions of survival. There are many other factors at play. For instance, conditions have to be manipulated so that the right atmospheric pressure is provided. Additionally, the gravitational force needs to be adjusted. The computer takes care of all the relevant factors. While robots can survive in different atmospheric pressures, humans cannot, so conditions are created to suit humans.

5
The Bermuda Triangle / The Physics of Space Travel III

An enduring mystery persists in provoking our attention and captivating our imagination. As scientific pessimism fails to temper our relentless instinct to suspect that these waters harbor a secret, the Bermuda Triangle not even whispers to us a clue on how she has claimed the many aircraft and sea vessels that have gone missing in her watch: not even lets a trace of wreckage out of her midst to allow the pessimistic to dismiss her from her throne of intrigue.

To those who are unaware, the region this book is referring to is also known as the "Devil's Triangle," and it is located in a stretch of waters in the Northwestern Atlantic Ocean. It is in these waters that a large number of aircraft and sea vessels have gone missing in mysterious circumstances. Strange anomalies are said to have occurred in this region, such as compasses spinning wildly out of control. This chapter will explain not just this anomaly, but also the fate of those who have mysteriously disappeared from this region.

Let us proceed directly to the heart of the subject: the Bermuda Triangle is the location of what is called a "fourth dimensional time tunnel." The Atlanteans possess the knowhow and technology to create shortcuts through space. Some may interpret this technology as a manipulation of time; it is a manipulation of speed.

In exploring the universe, it was long ago that the Atlanteans came across this sector of the universe. In exploring this sector, they came upon our planet. There is much to the history, but another book explains all of this. Notwithstanding, there came a time that the Atlanteans saw us from a new perspective and created a direct link from our planet to their home planet – that is, they created a fourth dimensional time tunnel, which became a shortcut between our planets.

Often exploration vessels establish fourth dimensional time tunnels on planets as a springboard to distant regions of the universe. The dimensional time tunnels link back to their home planet. The Atlanteans travel to distant and unexplored regions of the universe and along the way establish fourth dimensional time tunnels. They cannot create a direct link from their home planet to another planet. They must first go to the target planet and then create the link back to their planet. Once a link is created, there exists a permanent bridge, or doorway, between the two planets.

Once created, these fourth dimensional time tunnels will become permanent shortcuts for space vessels to travel to distant regions of the universe. There are many fourth dimensional time tunnels scattered all throughout the universe, and each one links up to the home planet of Atlantis. Space vessels are still exploring the universe as you read this, and new links are always being made. The universe is so vast, and the Atlanteans have barely scratched the surface of this universe, let alone any of the other universes.

When was the fourth dimensional time tunnel on this planet built? More than twelve thousand years ago. This fact alone changes the historical timeline that conventional historians and scientists recognize. This is not a subject for this book. *The First Cause, Volume II*, provides all the details of our ancient past, such as how the Atlanteans came to settle upon our planet, and how it was marked for evolution. This involves a deep subject on the hierarchy of the universe. This was an exciting chapter of our history, and it involved the origin of the pyramids around the world, the great flood, and so much more.

The question then arises of how this relates to the aircraft and sea vessels that have gone missing in the area known as the Bermuda Triangle. The answer is that they were caught up in the fourth dimensional time tunnel. The only instance that the fourth dimensional time tunnel was able to have an effect on aircraft and sea vessels in the area was when it was active.

The following is an account of what used to occur when the fourth dimensional time tunnel was active. The past tense is used here because, in our recent history, the Atlanteans changed some of the dynamics of the fourth dimensional time tunnel; one thing is certain, aircraft and sea vessels are no longer likely to become caught up in it when it is active. There used to be (and there may still be) an emission of green light from the technology involved, and a vortex. Additionally, the vortex region used to experience altered weather conditions, which abated when the vortex ceased. Anything in those skies or in that region of sea at the time of the vortex may have been drawn into it. Anything drawn into it traveled through the fourth dimensional time tunnel, and subsequently suffered the fate of being eliminated in flight through it.

Even though those who were unfortunate to have been caught up by accident in the fourth dimensional time tunnel did not physically survive the encounter, this is not to mean the story ends here. Nor is it to mean that the story is tragic in any way. We should all want for such a fate. The Atlanteans were able to detect their presence in the fourth dimensional time tunnel and rescue their souls. Their souls were then taken into a special room, where they went through the process of acquiring a cybernetic organism, as I did when I went to the mother ship when I was a young man.

Every single person lost to the Bermuda Triangle decided to stay on Atlantis and live for the term of 1500 years. They are presently a part of the Atlantean society. Ask yourself the question: Would you go back to our world if you were given the option to stay in their world? Every one of them is now employed, and has the opportunity to learn – and there certainly is plenty to learn. You may say that as their deaths were an accident caused by the Atlanteans, they were given a new life and are being looked after.

The fourth dimensional time tunnel was used extensively when the construction of the colony in the Pacific Ocean commenced. It was also used to transport the colonists to the colony. The Bermuda Triangle is linked to the mother ship, so that a spaceship does not appear in the vicinity of the Bermuda Triangle, but goes through a link directly to a dock on the mother ship, in the control center. When passengers arrive on the mother ship, they disembark from the flying saucer and automatically transfer to the living center.

Now, this book has tried to delve into the physics of much of the technology because its author has not been satisfied just to know that such technology exists. This is why this book will tackle the physics (from a layman's perspective) of the fourth dimensional time tunnel. By a process of logical reasoning, we should be able to come up with its blueprint, or, at the very least, a viable theory of it. Its very title gives us clues to its blueprint.

First, let us look at an idea once proposed by humans to replace conventional airborne flight. As we know, in the atmosphere, planes have to counter aerodynamic forces such as gravity and drag (air resistance). Once a plane no longer has such forces to counter, its travel capabilities increase. The theory proposed is to fly an aircraft directly into space, where it will travel across a region of space. Once it has reached a position in space that is relevant to its target destination on Earth, it will then reenter the atmosphere and descend to the target destination. Accordingly, if you want to travel from Australia to the United States, you could do this in just a fraction of the time it presently takes to travel by conventional means. Frankly, the only time consuming aspect of the flight would be the time that the plane leaves and to a lesser extent reenters the atmosphere.

In theory, this is exactly what the Atlanteans are doing every time they use the fourth dimensional time tunnel. Let us consider a flying saucer as the equivalent of a plane, the universe as the equivalent of our atmosphere, and the fourth dimensional time tunnel as the equivalent of space beyond our atmosphere.

What occurs is that the spaceship leaves the source dimension by way of the fourth dimensional time tunnel, it enters the fourth dimension, and then it reappears in the source dimension at the destination planet. This occurs in a short timeframe. It has bypassed who knows how many light years of space that it may have taken to travel by conventional means.

The line of reasoning is that once you have entered the fourth dimensional time tunnel you have left your source dimension. You have created a new dimension based on the speed signature at which you are traveling. Using the fourth dimensional time tunnel means that you are traveling at the speed signature that has been set for it. You are going to travel through the fourth dimensional time tunnel to the targeted entry point of your destination. To go through the entry point back to your dimension, you have to slow your cycle of speed to match the speed signature of the target dimension. It is that simple. Physicists may have a bit of fun figuring out the logistics and mathematics of all this, but this is the theory, and it is the future of space travel. All we need is the right technology and knowhow. For physicists, the logistics should not be difficult, because this book and the other books mentioned provide them with not just a starting point from which to base their research, but a direction in which to head.

This brings up many profound questions. What it tells us is that the fourth dimensional time tunnel has nothing to do with time, apart from the fact that it has saved time. Everything about the fourth dimensional time tunnel is about going out of one dimensional cycle and entering into another dimensional cycle. A dimension, which this books calls "empty," is added to the three dimensions of our world, and thus it is attributed the title, "the fourth dimension."

Time, as interesting a concept as it may be, is just that – a concept that is not tangible, just as a formula is not tangible. Some are of the view that time is a fourth dimension (this is attributed to Einstein), but this does not sound right. Neither does the thinking that time is interlinked with space itself, as in a space-time fabric (space-time is also attributed to Einstein). This book would go as far as saying that the concept of a space-time fabric as proposed by Einstein is inaccurate, whereas the model proposed by Newton has the right foundation. Newton believed that gravity is a force; that space is virtually empty; and that gravity is the result of forces of attraction. Einstein rationalized that gravity is not a force but a result of space being a fabric, interlinked with time, which he called a space-time continuum. In his model, the mass of an object distorts the space-time fabric: an object of large mass will distort the space-time fabric to a greater degree than will an object of smaller mass. Consequently, the object of smaller mass will be drawn to the object of larger mass.

As in Newton's model, gravity is the result of forces of attraction in the universe. It has a commonality with the forces of magnetism, although the two are different forces. To understand gravity, one must first understand the existing state of the universe and its origin, which means one must have read *The First Cause, Volume I*. For this reason, this book will not go into details on this interesting subject. *The First Cause, Volume II* will.

No one on Earth understands gravity at present. How can anyone? Much of modern day cosmology is built on foundations that no doubt will collapse after reading the above-mentioned books. Once scientists overcome the astronomical hurdle of allowing these foundations to collapse, a door will open for them to understand gravity . . . what a huge leap forward that will be for man! Understanding gravity will revolutionize the way we live; it will change our world for ever, from which there will be no turning back. On a philosophical level, perhaps this is why gravity is so elusive to us; perhaps until now we have not been ready for the changes that the applications of this one word will bring to our world.

Finally, going back to space and time, it makes sense to think of space as being interlinked with speed. The cycle of speed at which the atomic particles of the universe are traveling determines the space we occupy. How time fits into all of this is intriguing and puzzling to us, because it brings up relevant topics such as time dilation. (Time dilation is the relativistic slowing of time. That is, the clock of a space traveler moves slower than the clock of an observer back at the home planet.) Our ideas of traveling to the future and time dilation do not make sense. Even our concept of the future and past does not make sense. These are deep subjects, and are also discussed in the above-mentioned books.

There is not one universe; there are a number of dimensional universes, and they are populated in the same way that this universe is. However, there are possibilities of other dimensions that do not exist. These can be created. How we can look upon this is by comparing dimensions to a radio station and radio frequencies. There are frequencies that are in use by radio stations. If you want to create a new radio station, then you need to find a new frequency. This is all a dimension is: it is a "frequency" with a certain speed signature, with speed directly related to the cycle of speed at which the atomic particles are traveling. (This is also explained in the above-mentioned books.)

In other words, to arrive at the fourth dimension, we need to travel at the speed signature at which it has been set, which means that we need to travel at speeds that exceed the speed of light.

One could accurately assume that the fourth dimensional time tunnel is not just a door to a destination planet, but some type of accelerator that is able to accelerate a spacecraft once the spacecraft reaches the fourth dimensional time tunnel's speed signature. The spacecraft will initially require some form of propulsion to reach that speed signature. Nuclear fusion boosters provide the propulsion, and cold fusion is the nuclear energy provider. These nuclear fusion boosters are called "thrusters." Once at the predetermined speed, the spacecraft will have entered into the fourth dimensional time tunnel. Once in it, the spaceship no longer relies on thrusters to propel it; hence, this power supply is significantly decreased, or perhaps even shut down. The fourth dimensional time tunnel takes over and provides the acceleration required to propel the spacecraft. If you leave the thrusters on, you may overshoot the speed signature of the fourth dimensional time tunnel and create a different circumstance, which is undesirable.

In a crude way, you can compare a rifle barrel to a time tunnel, in that the rifle's barrel imparts a spin to the bullet. The fourth dimensional time tunnel imparts a spin to the spacecraft. There are no forces present in the fourth dimensional time tunnel to slow down or stop the speed of the spacecraft. Only the spacecraft itself can do this. When approaching the destination point, to exit the fourth dimensional time tunnel and reenter the destination dimension, the spacecraft must initiate a process of decelerating its speed to match the speed signature of the destination dimension. The spacecraft does this by hitting the so-called brakes.

One would rightly conclude that there would have to be physical structures on the two planets that are linked by the fourth dimensional time tunnel. This exists on our planet somewhere on the seabed of the Bermuda Triangle. When the fourth dimensional time tunnel is activated, this link establishes the target destination of any spaceship that enters the fourth dimensional time tunnel. This is important because it would be easy to overshoot your destination to the extent that you could wind up who knows where – perhaps lost in an unexplored region of the universe somewhere at the other end of who knows which universe. In some respects, going through a fourth dimensional time tunnel is in principle similar to traveling in automatic flight mode. You could even say that the links on each planet are to a spaceship what a lighthouse is to a ship at sea.

Fourth dimensional time tunnels would use a universal frequency. This frequency is the set speed signature. To enter the fourth dimension, you have to be able to travel at that speed. When you travel at that speed, you create the dimension. The dimension does not exist in its own right in the same way that the other dimensions of universes presently populated with matter exist. An empty dimension can only exist when you create it by traveling at its speed signature. Just for the sake of hypothetical analysis, you can create numerous other dimensions by traveling at different speeds.

Once you travel at the speed signature of the fourth dimension, then you will have created a three-dimensional vacuum. It is not the same as space as we know it, because even though outer space may look vast, it is not a vacuum. The fourth dimension is a three-dimensional vacuum. The dimension only exists for the spaceship that is in it.

We have now solved the mystery of the Bermuda Triangle: It is merely a shortcut through space, one that facilitates the travel needs of a superior race of humans, who just happen to be settled on this planet on two fronts.

6
Construction of the Colony

It took me a while to ask the most important question of all, "Where do you come from?" He told me that they were from a planet in the Orion constellation.

"When did you start all of this?"

"Around twelve thousand years ago. Initially, we had to bring all the required materials and personnel here. Scores of specialists were brought here. Initially, they came by means of leaping; then when they arrived, they all participated in doing what had to be done.

"In the first stage of construction there was a special room built. This room projects a green light from time to time, and it has a special purpose: it is designed to make mass copies of the human body." This is similar to the process I experienced on the mother ship when I was given a body. The difference is that this is equipped for volume, while the technology on the mother ship is not. That is, this one can mass-produce many bodies at once.

"Each time a spaceship designed to carry cargo came with raw materials for construction purposes, it also brought hundreds of workers – amongst whom were specialists, such as engineers. To allow the ship to carry ready-made framework for the structure of the colony, humans were not transported in their bodies. This alleviated some of the weight of the flying saucer when it arrived on planet Terra itself. Humans traveled metaphysically, compressed in several bodies (which is known as leaping). They were compressed not in a physical human body, but in a robot. This type of robot does not possess a soul, but he carries numerous souls, none of which belongs to him. This is a special robot designed for this purpose, which means that he is equipped to carry many metaphysical bodies.

"When many souls are compressed in a single body in this way, they still feel conscious and are able to communicate with one another telepathically. However, the robot housing the souls is always in charge.

"Once on the planet, the metaphysical bodies were separated from the robot and given a body. This occurred in the room that gives off a green light. How the room processed these people went like this: one person went in one door and a large number of people came out from the other door. The souls were one by one detaching from one body and reattaching to a new body, until every single soul came out as an individual with a body of his own. They were not given their individual appearances; they were clones, and their bodies were cybernetic organisms.

"They only occupied these bodies while the project was underway. When the project was completed, either they went home by the same process that they came here, back to their own bodies, or their bodies were brought here to them. In this case, they may have joined the colony as a permanent resident. These specialists are regularly sent to planets for just the same purpose: to construct a new colony or to expand an existing colony. Your planet is not the only planet we are colonizing in this way.

"It took a number of trips before all the technology was brought here, not to mention the thousands of engineers and specialists. Ever since, we have continued to expand the place.

"The first thing we did was generate an anti-magnetic field, combined with gravitational forces, in the water to pressurize the place – most of the water was just pushed out of the area in which we planned to build. In building the colony, we were responsible for causing your sea levels to rise. To enable such a structure to be built, we had to remove and cut down a lot of rocks and mountains to make the ocean floor level. We used particle technology to do this. Particle technology converted the mountains and rocks into subatomic particles. We didn't have to do any actual lifting or cutting, and there was no waste. Then what we did was use laser technology to level out the surface.

"Once we built the initial stage, we expanded it further and made it bigger in size. It continues to grow in size as the colony expands.

"The same technology was used to build the pyramids. Many specialists were required to build them, as well as many workers who could operate the technology."

Now we can appreciate how such a structure could be built under the ocean. They did not need to have special underwater vessels to build it – let alone have several million Egyptians fill their lungs with air and then hold their breaths as they dragged tons of rocks onto the ocean floor to build it! They simply turned the area into an oxygen-rich environment, which was maintained on the ocean floor by a combination of gravitational forces, antimagnetic forces, and some other forces of which I am not aware. These forces created a force-field effect around the large so-called pocket of air to keep the water out and the oxygen in. This meant that they could walk on the ocean floor as though they were walking on land. They could breathe air and live under the ocean, just by the use of sophisticated technology, and by the creation of environmental conditions that support human life. The fishing areas described in an earlier chapter were created in just the same way. They have no obvious barriers to divide the ocean from the pocket of air that they use to catch fish. Each time a section is added to the colony as it expands, they use the same method cited earlier.

This is a basic summary of how they constructed the colony, as I understood it. When you come to think of it, it would certainly have been a challenge to construct it in any other way. They have ingeniously found the simplest way to construct an engineering marvel.

7
Farms in the Pacific Ocean / Agricultural Science / Tablet Manufacturing / The Physics of the Food Tablet

One of the screens showed a beautiful field of crop that could have been wheat. The artificial light, which gives the impression of sunrays, and which contains the essential ingredients for agricultural development, was lightly touching fields of crop that seemed to go on for as far as the eye could see. The undulation of the crop suggested a gentle wind, which surprised me. I was told that they create not only wind, but also rain in these fields. Rainfall is thus a method by which they water crops. Artificial humidity is created, which is the seed for storms, complete with thunder and rain.

The land on the screen before me was mainly flat tablelands. There were some trees in the distance. I was feeling both peaceful and weird: peaceful in knowing that I was seeing something that was so familiar to me on the land: weird in knowing that I was seeing the same landscape on the seabed. Weird also in knowing that you are capable of walking in those fields and breathing in the air.

On one screen the fields just kept coming at me, as though I were airborne and flying over them. There never seemed to be an end.

"How big is all this?"

One of the men smiled and said, "These are our farms, and there are miles and miles of them. The farms that you see are contained in a separate region of their own, at a distance from the colony.

"We are able to genetically modify any crop to grow in any condition that suits us. Once you genetically alter a crop to suit prescribed conditions, and then you reinstate that crop in its native conditions, it will no longer survive in those native conditions."

On another screen I saw a three-level complex. Each floor of this complex has high ceilings. What is interesting is that in this structure not only are the fields of crops grown, not only are the crops stored, but the food tablets are manufactured and stored. This means that the food that is grown on the farms is not just stored, but also converted into tablets, in this structure.

The ground level is the farming level, and it is here that varieties of crops are grown. It is on the middle level that crops are stored and dried before they are converted into a tablet form. It is on the top level that the conversion of food into tablets occurs. The technology that is used is mind blowing; yet what appears to be so complex is so simple in theory. The top level is also a storage region for the finished product – that is, the tablets. As a finished product, tablets occupy a fraction of the total space; most of the top floor is used for the production of tablets.

It is here, on the top floor, that tablets are appropriately labeled, packaged, and then stored. Boxes are placed on platforms that hover. There are no wheels in their society. They have done away with one of man's greatest inventions. They have the technology of the gods themselves. This is why their civilization is run by some of the consciousnesses of what some of us call gods.

On another screen I saw how they harvest crops. The crop I saw was wheat. Hovering above the crop at a slow speed was a wheat harvester. Following it was a second wheat harvester. It was leveling off the top of the wheat and collecting what had been missed by the first wheat harvester. There were over two dozen flying saucers operating at once.

The following is an account of how the harvesting machinery works. It has a large "arm" attached to it. The arm is outstretched, and it cuts and collects the top half of the crop. The cut crop is drawn into the arm, which is large enough to store a large quantity of wheat. In here, seeds are separated from waste. The seeds are collected, while the waste is converted into powder, then blown out, and evenly dispersed over an area of cut crop. When it falls on the ground it acts as a fertilizer. The bottom uncut portion of wheat is not removed.

Once the waste has been evenly spread over the fields, another piece of farm machinery that uses antigravity technology hovers over the half-cut crop stems and reinvigorates them. This means that the stem will not create side shoots. The same shoot that was cut grows back again to form a new single stem and head of wheat. This reminded me on the single cell nature of a worm, which, when cut up, will grow into a new worm. Moreover, this harvesting practice is applicable to the harvesting of all the crops.

I saw apple trees being harvested. I figured out that there are no apple pickers in their society. The farm machinery chops the trunks of apple trees so that the leaves and branches are completely removed from the trees. All that remains of an apple tree is a small portion of the trunk. A second harvester then collects any missed branches.

The arm of the harvester separates the leaves and branches from the apples. Approximately seventy percent of the waste will become water because of the high water content of apple trees. The remaining thirty percent residual waste is turned into powder. Both the powder waste and the water are evenly dispersed across the fields.

As in the case of the wheat crop, the stems of the apple trees are reinvigorated so that the trunk grows into a tree again, and goes through the same cycle. This regrowth of not just the apple trees but also all the crops and fruit trees, is, in my estimation, a two month process. This means that after being cut, the crop or the fruit trees will be ready for harvesting approximately two months later.

Clearly, their crops are genetically modified for a rapid cycle of growth. As applicable, they are also loaded with vitamins, minerals, and protein. This enhances the flavor of the produce.

The collected fruit is taken to the middle floor of the structure, and treated as it needs to be treated. Some requires drying. I would estimate that the quantity of food I saw stored there would be enough to feed the colony for five hundred years.

When the process of conversion from the original product to the virtual equivalent of the product occurs, each item of produce is processed separately. This means that each has its own process of conversion. Although there are no conveyor belts, as we know them, there is a sophisticated form of this technology. Apples will go along one conveyor belt, while wheat will go along a different conveyor belt.

A conveyor belt takes each product to a particle converter, which converts the product into subatomic particles. Once broken down into subatomic particles, the product's virtual equivalent is then registered in a library. For instance, when wheat is broken down into subatomic particles, virtual wheat will be registered in a library. The same with corn. For that matter, this applies to everything that is converted into subatomic particles. In other words, anything that is broken down into subatomic particles is registered, and this documentation occurs in a library in the form of a virtual equivalent of whatever it was that was broken down.

Now we arrive at the next stage of the process: the manufacture of food tablets. Let us consider the case of a bacon and eggs tablet, which converts into a dish of three rashers of bacon, with three sunny-side-up fried eggs, and half a grilled tomato. We must consider that as a part of this tablet there is also a plate under the food, along with the spices that flavor the food, such as salt and pepper. It may just wrack our brains trying to figure out how one tablet can convert into a meal with all these ingredients, perfectly arranged on a plate. It seems impossible, but not so! When we consider the tablet in terms of subatomic particles, the simplicity and beauty of the technology becomes apparent.

First, we must understand that everything has its own unique arrangement of atoms. (Atoms, as we know, are made up of subatomic particles.) Take yourself: like everyone else, you are a unique arrangement of atoms, which is why we all look different. If we take a boiled egg, it will have a different arrangement of atoms from a fried egg. A hot boiled egg will have a different arrangement of atoms from a cold boiled egg. This suggests to us that there is another factor to consider: temperature. Thus, heat is one of many variables in the formula for a cooked egg.

The simplicity of putting together a food tablet in the manufacturing stage involves the subatomic particle formula, or blueprint, of each item. Once you know the subatomic particle formula of something, creating it from subatomic particles is easy. In the developmental stages of this technology, no doubt, much research was done in formulating the blueprint of every single product there is, including a bacon and eggs dish. One rasher of bacon has a subatomic particle formula. Two rashers of bacon will have a different subatomic particle formula. Crispy bacon has a different subatomic particle arrangement from lightly grilled bacon. I think we get the picture. What is relevant here is the subatomic particle formula of every single product, with each product having a unique arrangement of subatomic particles.

This means that the subatomic particle formula for three fried eggs, three grilled bacon rashers, half a grilled tomato, flavors, and one plate needs to be a part of a tablet to make a bacon and eggs meal. Imagine having a perfect dish of bacon and eggs on the table before you, ready to eat. If you could break down that entire meal into its subatomic particle state, then that is the formula you must have on the tablet.

With this in mind, the technology that creates the food tablets has a computerized program that contains all the blueprints – that is, all the subatomic particle formulas of every product in existence. Consequently, when a tablet of bacon and eggs is being created, the program knows the subatomic particle formula for a bacon and eggs dish.

It is in the manufacture of tablets that the virtual library comes into play. This library, if we recall, is a virtual register of products that have been broken down into subatomic particles. A food item in the library is the virtual equivalent of a real food item. Therefore, you cannot create a tablet when the virtual raw products are not registered in the library, just as we cannot manufacture a product if we do not have the raw products in stock. Let us take a hypothetical scenario. There is a shortage of eggs, and the register of virtual eggs in the library has run out. What will happen then is that the bacon and eggs dish cannot be produced. Virtual eggs will have to be outsourced from a trading partner if the colony is not self-sufficient to provide the matter eggs to restock the virtual eggs in the library.

Let us take a step forward. When the food tablet goes into your food converter in the kitchen, the food converter needs to know what it is going to create. The tablet itself offers no clue to what its outcome will be unless there is some form of information on the tablet. This information is the subatomic particle blueprint of the meal. Only with this information can the food converter convert the tablet into a meal, such as a bacon and eggs dish. This means that the tablet is only a token of a product.

If we go back to the manufacturing stage of a tablet, information is provided on the label of the product. This could even be in the form of a miniature chip that would go on the packaging.

To convert the purchased product into a meal, the food tablet will go into the food converter that you have in your kitchen. The food converter then does all the work. It reads the formula for the arrangement of the subatomic particles, and then it arranges subatomic particles according to this information. When it does this, the food will be created to perfection at the required temperature. There is no such thing as overdone eggs. This is because the mathematical formula for overdone eggs has not been supplied. The product will always come out perfect from the food converter, according to the specifications on the food tablet.

Accordingly, we should now understand that a tablet is merely a vessel that contains the mathematical formula for the arrangement of subatomic particles, and this arrangement will produce a matter product – such as a bacon and eggs dish, complete with the plate. Obviously, the mathematical formula for the plate is provided. It is simple, is it not?

Whereas the food converter is the technology, the chip containing the blueprint is the brain, so to speak. Having the technology to arrange subatomic particles in a particular manner is not as complex as it seems, once you have mastered quantum mechanics and understand the nature of subatomic particles, which at present we haven't and don't respectively.

In summary, once a crop is converted into a subatomic particle state, a count of it is recorded in a library, which is a virtual replacement of the actual product. In the creation of food tablets, stock is accessed from the virtual library. If any of those virtual items have run out, then it is not until they are available that food tablets can be created. There are strict controls on this technology. Programmed in the technology that creates food tablets are the subatomic particle signatures of every meal and product there is. The tablet is packaged with a chip that contains the subatomic particle blueprint of the meal. The food converter in your home reads the blueprint and arranges subatomic particles into the relevant meal. Clearly, the packaging disappears in the process, along with the chip. Moreover, these must convert to subatomic particles in a self-termination process.

Finally, we may have asked ourselves the question: Why does a tablet require refrigeration, when it is nothing more than an information vessel, which has no obvious perishable properties? At first glance this may seem so; however, the basic idea of food having a perishable quality has been applied to the food tablet. The tablet is the virtual equivalent of a food dish or a food item, and it has perishable properties, in the same way that the non-virtual equivalent has. This means that if an item is left in the fridge for too long it expires and needs to be thrown out. The food tablet has intentionally been given a perishable property; for this reason, it requires refrigeration. Why it has been given this is for the sake of the economy. More on why an economy is important is explained in the next chapter.

8
Economy and Trade / Legal System

The manufacturers of food tablets, as in the case of all manufacturers, are privately owned enterprises. The corporate bodies involved have guidelines that have to be followed. These guidelines strictly prohibit the use of the technology in any way other than those set out by the economic administrators, who are the equivalent of one arm of a government bureaucracy, only they work under the auspices of the governors. All technology is controlled. Certain technologies are not allowed in the hands of the population for personal use. Guidelines exist because the basis of economies could be in question.

Even though they have the technology to produce food from the building blocks of subatomic particles, which literally means from nothing, they are not given the freedom to produce products from nothing. They need to grow corn to make corn products. They need cows from which to create steak tablets. They need chickens to lay eggs to make egg products. If a manufacturer does not produce a product that it requires, then that manufacturer will have to buy it from a trading partner. Businesses have established trade partners and manufacturers from which they can purchase any material or any product they need. And there is a thriving trade going on among many civilizations.

The idea is that you must have economies in the universe. There is a golden rule in an advanced society: technology cannot replace an economy. People have to work. Technology can be a trap for an evolving civilization to revert to a stagnating civilization. The phase-1 human is a classic reminder of a stagnant civilization of humans. (An account of the phase-1 human exists in *The First Cause, Volume I*.) Overwhelmingly, to get a person to use his brain is something that has to be forced upon him, whether he realizes it or not; education and work are two such methods. To maintain an economy, it will always be the case that some technology is strictly controlled and unavailable to the public.

Just to illustrate how controlled subatomic particle converters are, and how strict the use of subatomic particle converters is, let us consider the case of leather. We already know that leather can be created from the breakdown of the skin of some species of fish, mainly from larger species of fish. Once skin that has the properties of leather is broken down into subatomic particles, the virtual equivalent of it can be used to put together the building blocks of leather. The virtual fish skin cannot be used to create cotton. It is in the interests of the economy that a virtual product is used as if it were a matter product. If there is a shortage of any one virtual product, then that product must be purchased, either in its matter form or in its virtual form. This means that there is a thriving trade of virtual commodities. One would expect that in the interests of economies, virtual commodities have shelf lives just as matter commodities have.

On the mother ship or in the colony, there are no cows from which to source steak tablets. Companies import virtual steaks from a trading partner. This means that the manufacturer of a steak, chips, and gravy tablet will outsource its steak from a wholesale meat supplier. From this butcher, a manufacturer will buy virtual cuts of meat to suit the servings of the steak, chips, and gravy meal. Let us assume rump steak is the preferred cut of meat in the meal. Then virtual rump steaks will be purchased. Once the manufacturer has the virtual steaks, then the steak, chips, and gravy tablet can be produced and sent to the supermarket for sale to the public.

Money is an integral part of any society. Cash and currencies exist. Just as there is a food converter, there is a money converter. No matter how evolved a society, you will always find those who still want to see and count cash. People still carry wallets. Even dolphins are aware of money and its value! This is another instance of why technology is highly guarded: in the wrong hands, it could spell ruin and chaos. The advantage is that robots and even computers know your thoughts. There is next to no possibility of cheating or corruption in their society.

・ﾞ

When you start a business in their society, you need to register that business, in the same way that we register a business. When it comes to the high end of business, as opposed to small business, companies are given the equivalent of a ring. This is no ordinary ring, and it is certainly not worn as an adornment. An administrator, who is a robot, gives the ring to you. Each board member may also be given a ring. All the "big shots" of a company – that is, those in the decision-making process who have "make or break" powers – are given a ring. When there is a board meeting, everyone usually wears his ring.

The ring has an identification code that is unique to each company. Therefore, when more than one ring is given to a company, each ring carries the same unique identification code. How to identify the ownership of a ring is by the color signature. The color spectrum is so vast that no two colors are ever the same. The beautiful thing about using color signatures as a source of identification is that they are as unique as DNA prints.

When you combine the color spectrum with technology, its potential is far-reaching. Once a company is given a color signature, it can never be copied or replicated. Trying to copy a color signature is the same as trying to copy someone's fingerprints, or DNA print.

Besides having a unique color signature, a ring possesses robotic technology. Robotic intelligence in the ring is like a spy in the sense that it records all the data relevant to every transaction that goes on in the company. Additionally, it records the data of everything that everyone in the company does (on a business level), particularly of those who can make or break the company.

The ring cannot be lost or destroyed, and decisions or transactions can never be disputed. In a way, this is like a tracking device, only it tracks everyone's moves, actions, and thoughts in relation to the business. It screens out anything irrelevant to the company. The ring is even capable of producing a holographic projection of the day-to-day actions of every member of the company. The information that is stored on the ring cannot be tampered with.

In our society, far too many companies go bankrupt after stealing money, often from the little man, who means nothing to them. In the Atlantean society, such a thing can never happen. Despite whatever we know about the moral nature of Atlanteans, there will always be the rare few who will come up with something devious based on greed, and this is because they have allowed their negative side to prevail.

Let us not be fooled, companies can go broke in their society. In such an instance, the equivalent of an administration team steps in. The administrator is a robot, who acts under the authority of the governors. He will take possession of the company's ring or rings. The ring is a vital record of all the transactions made by the employees and those on the board of directors. The ring is also a money trail. This means that there is no way anyone can ever hide or steal money from the company.

The ring or rings never have to be worn by anyone in the company. Just because a ring is not present in the back rooms when deals are done, it does not mean it is not recording the deals. On the contrary, the nanorobot associated with the ring or rings is in telepathic contact with everyone involved in the company. There is no way someone can hide money without the nanorobot knowing. Someone can even deliberately throw the ring away. The ring will merely telepathically communicate its fate to the administrator. You cannot destroy information once it has been collected on a telepathic basis by a nanorobot. Every nanorobot is connected to, and is a part of, the conscious computer; if we take a hypothetical scenario of a nanorobot being lost or destroyed, its data will have already been logged in the conscious computer. This ensures that those who are capable of destroying the economy, and invariably society, are incapable of such destruction.

We must remember that, in their society, large corporations do not make the kind of profits our large corporations make. Once a company's income reaches a predetermined limit, the remaining income is payable to the governing authority. This is why there is no rich or poor in their society. What is happening in our society is unconscionable. We have people looking for food in rubbish bins; we have people struggling to find a job to pay the rent or mortgage; we even have people whose income is not or barely enough to pay the rent or mortgage, let alone the living expenses. Many battlers lose their homes because of difficult times, through no fault of their own. Many are in dire circumstances, and there is no support provided to them. All the while, there are those who are paid obscene amounts of money.

·ب·

If someone is subject to prosecution, for whatever the reason, there is a court of law. Presiding in this court are three judges, all of whom are robots. Now then, you may wonder, who goes to court in this near-perfect society? This is easy to answer. Anyone who has committed murder is just one example. Such a human is rare, and is usually caught in the planning process, but there might be an occasion that someone might not be detected in the thought process. By the time someone goes to face the judges, he is already guilty. An innocent person will not progress to this stage.

Anyone brought before the three judges will wish he were never born. He knows what fate awaits him. A convicted person is sent into a special glass box. Once he is inside it, a blue fog appears and within seconds he becomes particles of dust. His remnants are then sucked out through a suction system and spit out into the ocean to become a source of fodder for marine life. His metaphysical body is not affected; it goes through the cycle of death, in which it faces another round of judgment. What this process involves is explained in another book.

The courthouse itself is just an office. There are no juries. There are only three robots who read your mind; they don't allow you the opportunity to say a word in your defense. You have no lawyer. You have no avenues of appeal. You have absolutely no rights. For just a few seconds you find yourself in front of a panel of three. Judgment is swift; with judgment, you are instantly transferred into a locked glass box. In those seconds between seeing your judges and appearing in the glass box, you feel a total sense of panic. That panic plays havoc with your body, and you experience new sensations. For one thing, there is no blue fog to convert your involuntary release of refuse from your body. For the first time in your life, you discover what feces looks and even smells like. Your final breath is not a sweet smelling one! All you hear is a small noise, and then you are gone. Nothing of you remains except dust. Your soul is still there in that glass box, and it sees your dust sucked out through a chamber. No trace of your physical existence remains; all that remains is the unpleasant memory that lingers in your intellect, which you will carry with you when you face another round of judgment. Your soul in this instance dreads ever having created such a memory.

9

The People

One of the men answered a thought in my mind, "We often visit the surface of your planet and mingle with you. We do this in a physical way."

"Why would you do that when you have everything here?"

I was not expecting his answer. "For the same reason you go to visit the zoo. You entertain us. Sometimes we fail to understand your logic."

I understood what he meant by this completely, and did not take offence, but agreed with him.

"This colony here in the ocean is colonized by families. Many individuals have chosen not to live down here but to live on the top with you. You just don't know who they are. They are forbidden to say anything. Many of our people are your teachers, and they can be found in many fields. When they go to live on the surface, they don't even know that they are from Atlantis. One thing about them is that they are achievers, and they all have a high IQ. They just help advance your civilization by way of influence, and by way of the introduction of new technology. Ideas you see in your movies have been suggested by us and are based on a reality somewhere in the universe. It is a way of acclimatizing the population to accept the reality as it draws near, rather than hitting them in one go as Jesus did. Mentally, we are in tune with these people and their minds are wiped if they are about to reveal knowledge that they are not supposed to, which is because it may be too early. Those who know who they are, who are in the right positions, such as in governments of the right countries, would not want a human to know who they are. Your people would probably kill them." This is an extensive subject, and more of what I came to learn is explained in another book.

If you think about it, to an advanced civilization, visiting us would be like traveling back in time. Watching our evolution would be a wonder of its own, but a completely new dimension would be participating in it, contributing to it, or just stepping into each stage of it as a tourist.

On one screen I could see children. I was told that there are numerous children in the colony and that it is not an old society where there is a small number of children. In response to my thoughts, the same man said, "People do have sex to populate our civilization. That is our primal urge. However, a superior human no longer has the kind of sex urge you have on your planet."

I knew then that unlike humans from our planet, these people have nothing to hide, which explains how couples can read each other's mind and still get on so well. On our planet, many humans are driven primarily by the sex urge, and it gets some of them into trouble. They also have to hide their bedroom antics and indiscretions. No one wants others to know what goes on behind closed doors; the problem of a superior race of humans is that telepathy allows one to know exactly what goes on behind closed doors, apart from that which goes on between a husband and wife. This is why it is an unfavorable thought for humans on Earth to read one another's minds. Privacy of the mind is jealously guarded and highly cherished by a primitive mind, because of all the "skeletons" contained in the "closet" of that mind. In advanced societies, people don't have secrets of this or any nature; they don't have indiscretions or skeletons. This way, reading one another's mind is irrelevant. They don't care about privacy because they don't have anything to hide.

One of the men explained, "The sex urge in a superior human ceases to exist, except if the species is a reproducing one, in which case it serves the purpose of reproduction. Somehow, the hormone factor just changes once they have two babies. As the sex urge is tied to reproduction, when the body no longer has the hormones required for reproduction, such as when you have had your two babies, then the sex urge declines. When you are young you have a sex urge, but after the kids come along, the hormones change. Even young ones don't think too much about sex. Their minds tend to be preoccupied with, and concentrated on, the acquisition of knowledge."

The other man said, "In being able to read minds, these people stay honest. This is one way to keep an honest society. This also means that there are no such things as perverted sexual behaviors, as you have on the surface. The negative side has influenced these. Here, everyone has traditional values, and there are no exceptions in an intelligent society."

On a television screen I noticed that kids and pets were running around. One said, "Children and pets alike can read your mind. When it comes to pets, they don't ever fight. Violent temperament has been completely removed from their DNA. As for humans, they have not attainted equilibrium yet (this is a deep subject, explained in full in *The First Cause, Volume I*), so they sometimes argue, but that argument is usually a hot debate, and that is as far as it gets. They don't get all heated up and punch each other as humans on the surface do, or exhibit physical violence of any kind.

"The kids have a lot of knowledge because they study every field of specialty, such as technology, physics, chemistry, mathematics, the arts, and history. They begin their education immediately. They don't waste time as you do. The baby's education begins as soon as it starts to crawl and play with toys."

One thing I could not see was a swing. I was amused when he said, "These kids are one hundred times more adult than humans are on the surface. You don't want a kid like this to tell you off telepathically. If they want to see or use a swing, they go to visit the 'zoo' on the surface. However, they don't find it particularly entertaining. They would have nothing in common with your children. If they did find your entertainment interesting, they would probably bring your primitive technology back with them here, but look around you, you don't see any swings here!"

10

The Streets, Homes, and Life in the Colony

Life in the colony is not different from life in the mother ship. The technology is all the same. The people are the same. The only difference is that the mother ship offers recreational facilities to the inhabitants of the colony, which the colony, largely, does not. I learned that the mother ship is destined to leave the Atlantic Ocean one day, but this will not be for quite some time. When she leaves it will be because her job will have been completed, and she will seek out a new adventure and a new challenge. By then, we should be a part of that adventure. Moreover, when the mother ship leaves, the colony will have already turned to the surface of our planet for its recreational facilities. By then, these people will have mixed with us because they will feel secure enough to mix with us openly. For this to happen, man will have to be drastically overhauled – his present state leaves much to be desired. We have to catch up with them intellectually, behaviorally, emotionally, and technologically. In particular, we need to discover the blue fog technology. Why we need the blue fog technology for them to live here is that living in any other environment will affect their ageing cycle, and make them prone to those negative aspects from which their environment makes them immune. They are not going to come to the surface to live in our environment until all these factors are addressed.

In one sense, having them come to live with us at our present state of intellectual development would be the same as having lambs mix with wolves. We need to control our negative side and be sincere and honest. My advice to the Atlanteans is that Chisek (the name of a negative force, which is explained in another book) is predominant in the best of humans. One does not need to do much to invoke Chisek in a human; knowing more than someone else is all it takes, or the perception that you can read someone's mind and know his skeletons. I would say to stay away from us, unless you want to live like that wise man, isolated on a hilltop. If you have any doubts about this advice, just step on a cat's tail and wait for the music! No doubt, you are intelligent enough for me not to have to make the connection between a cat and a human.

In some respects, the colony is much the same as a housing estate, and it is reliant on the mother ship. Every five minutes or so there is shuttle service that transports people to and from the mother ship, and the trip only takes a few minutes. This gives people access to amenities not available to them in the colony. In the colony there are limited recreational facilities. For instance, in the colony there are public swimming pools in each local area. There are around six pools in every swimming center. There is also a small park at these swimming pools. The public swimming pools are open twenty-four hours a day.

Dusk and dawn are recreated by the lighting system. Night is much the same as our night with a full moon. Their ceilings at night create a starry twinkling sky. In the evening, the effect is magical. There is no pitch-black effect of night, and there is no bright object in the sky, such as a moon, to negate the effect of a starry sky. During the day the ceiling is light blue in color, and it creates the impression of a sky.

The colony is fully air-conditioned, and this air-conditioning creates a breeze. When hot and cold air mix they form clouds, thunderstorms, wind, and rain. You can even see the build-up of clouds. You can tell by this build-up when there is going to be a storm.

This is a natural process of hot rising air meeting cold air-conditioned air. Two different air fronts meet and create storms. Even though this is a natural process, it must be said that there is a measure of control in these storms and rain. They are controlled to recreate a natural environment.

The colony is composed of housing estates, which are divided into local communities. The surface of some roads is an attractive olive-gray color, as are the footpaths. Some roads where relevant are gold in color. The roads are anything up to one hundred meters wide.

Running along every street are rows of double-storey houses that are appealing and different in design. Each house has its own small backyard, which is about twelve meters long. If you have kids, this is an ideal place for them to play. There are gardens in the yard, and each yard is landscaped a little differently. You have the option of choosing your home in which to live, and you choose the garden style that comes with the house. All the designs are exceptional. Even though you have a backyard and a garden, you will never do the gardening yourself. A robot goes from house to house and maintains all the gardens and landscaping. This robot also looks after the nature strip out the front of your home. This is similar to a hired gardener, only you don't pay for his services, just as you don't pay for your home.

At the front of every house is a porch that extends out to the public footpath. The porch runs across the width of the house. On the upstairs level there is a balcony, but it is smaller than the porch. You usually keep your collapsed shopping trolley on the porch. As we know, the shopping trolley can fold up to the size of a suitcase, as impossible as this may seem.

There are no front gardens; there are no swimming pools in backyards; there are no garages. In the colony, you do not own a flying saucer as you do if you live in the mother ship. To serve your transport needs there are public shuttles that act as buses. There is always a shuttle going around the colony. These all use antigravity technology. Your only option in the colony is to use public shuttles to go to work. If you have to go shopping you have your shopping trolley, and there are local shopping centers that service the local community.

The interior of the houses do not need a description. All rooms have a décor menu from which to choose a style of décor. Everything applicable to the interior of the unit in the mother ship is applicable to the interior of the houses in the colony, only the rooms are spread over two levels. The upper balcony and windows have an antigravity safety system operating outside them, which prevents anyone from ever being injured if he were to fall over an edge or out of a window. Indeed, wherever there is an opportunity to fall, this technology is used as a preventative measure.

The nature strip that runs between the street and the footpath has French-style lamps. There are usually three lamps on a single lamppost. On the nature strip there are trees and landscaped gardens. There are birds, but these are not the same as those on the mother ship, in the sense that they mind their own business and do not have conversations with you. They certainly don't intentionally leave droppings on your shoulder or head!

The reason birds in the colony are not as advanced as those on the mother ship is that all the life forms on the mother ship are old and established. That is, just like humans, they live long lives. When the life forms eventually die, their metaphysical components are reincarnated back on the mother ship. Their continual cycle of reincarnation on the mother ship explains why the native inhabitants have evolved intellectually as they have. The mother ship has seen many planets before us evolve. We are not the first planet the Atlanteans have colonized, and we will not be the last planet that they will help evolve. By contrast, the colony is young, and the birdlife is new to the colony. Just as the birdlife in the mother ship has taken a long time to reach the degree of intelligence it has, so the birds in the colony can look forward to a similar evolutional process. The intelligence of some life forms evolves rapidly when those life forms have repetitive reincarnations in the same location.

Between the nature strip and the footpath is a strip that is only about two inches wide. This strip looks as if it has thick glass on top. Whatever its composition, it is strong and never chips. It looks like a neon light, and it illuminates at night. The colors it emits alternate. The trees are laden not with fairy lights but with long diamonds. They dangle loosely from tree branches, and at night the neon light bounces from them. The breeze blows through them, and the multitude of colors they reflect looks brilliant, so much so that it is impossible to express the beauty of what you see and feel in words. It is analogous to trying to describe the feeling of that one moment in your life that exceeds all other moments.

When I was viewing this scene, I remember asking the two men a logical question, "What would happen if someone fills up his pockets with and steals some of these diamonds?"

Both smiled, looked at each other, and laughed aloud. One said, "First of all, the diamonds are going to gain weight, so that it will be practically impossible for him to even walk, let alone try to get out of this place. Second of all, it would be virtually impossible to sell a diamond to anyone here, because nobody here is interested in buying diamonds."

When you sit on your balcony and look up and down your street, you think you are in heaven. You do not have a worry in the world. You are as at peace with the world as the birds that sing their twilight tunes in those trees laden with diamonds. These humans live as nature intended them to live: as free as the birds, without the worry of having to put a roof over their heads or food in their mouths. One great desire motivates them: to evolve their intelligence.

Conclusion

It takes courage for one to speak of things that are outside the narrow tunnel that most of us see through. After reading this book, those who are honest with themselves will not be able to giggle at it, or dismiss it as a work of fiction. Before anyone thinks to tag this book with a fiction label, he should remember that even the idea of black holes was giggled at, and not all that long ago. Yet it is for some of those that we giggle at that our knowledge crawls forward.

Many are wondering about what the end of some ancient calendars means. Our forbears were told of the day that we would enter into a new phase of our evolution. This new phase marks the end of imposed darkness and ignorance, and the beginning of a new era of understanding of who we are (our blueprint), what happens to us after we die, and what is out there in the universe. Many profess to know answers to these profound questions of life, and yet . . . many words always leave us wondering more. By contrast, this book and *The First Cause, Volumes I and II* answer all of these questions.

As impossible and as far-fetched as it seems, we are the generation that is being presented with the truth. In the case of what is "out there," this is irrespective of governments and their cover-up mentality and propaganda. This goes beyond the governments. Governments, as with vested interests, will never tell us the truth.

Of course, there is no way we could know that it is only when a planet reaches a certain stage in its technology and invariably its intelligence, which presumably go hand in hand, that it is deemed ready to belong to the community that unquestionably does exist out there in the universe.

There is a higher order that exists in the universe, which evolved through evolutionary processes of the universe. Darwin thought on too small a scale; unfortunately, we still think small; then again, the truth is not something we can ever figure out, invent, or imagine on our own without some form of "external" intervention. The foundations we have built around us to explain away our existence, from Darwin's Theory, the Big Bang Theory, to spiritual and philosophical theories, do not form credible foundations upon which to structure the answers, just as the theories of the Earth being flat and at the center of the universe were not credible foundations and hence crumbled like castles of sand.

What we cannot imagine is that every planet in its primitive stage of evolution is allocated a superior race of humans to "oversee" it and aid it in its evolutional process in various ways. Our overseers have been "grooming us" throughout our process of evolution. The "footprints" they have left in our ancient past are all too evident today, if we look with open and unbiased minds. Believe it or not, much of our progress has been of their influence and not of our own innovation. Everything we think we are inventing is a reinvention of something others have long ago invented. The end stage of this grooming process is unification.

The question of "contact" is why many are dismissive of the idea that there is something more out there. However, it is not a matter of contact, as we have always been a "zoo" to others out there who have always known about us, and who have regularly visited us, unbeknown to us. Only when humans are ready, contact can occur in a way that will satisfy us, but it will be in a way that we cannot suspect.

Actual "contact" is not as simple as some may think. Contact on a global scale cannot just occur. There has to be an acclimatization process. As a part of the process, we must first understand not just ourselves, but who they are and where they are. This is where this book and the books earlier mentioned come into the picture. As far-fetched as this is going to seem, as a further part of this process, it has already been written that a group of 49 single humans from across the globe, who have long ago been chosen by the Atlanteans, will be physically taken on a tour (for several days only) of their civilization. These 49 people will be eager to go: one cameraman, one reporter, and forty-seven others. Our governments will neither know this in advance, nor know when. This book has described the very same tour that these 49 people will take. The purpose is for these 49 people to convince humans of the truth. These people will be instrumental in gearing up the world for the new phase

Conclusion

of our evolution, which will include further genetic engineering on a mass scale (unbeknown to us), to upgrade our intelligence. This is not as absurd as it sounds, and it is the subject of another book.

This "tour," by the way, will not be an isolated occurrence. Humans will regularly be chosen to visit them in groups for a period. Once humans are ready, contact of a different nature will occur.

Of this we can be certain: our planet can NEVER find peace on its own. Peace will only occur when our present governments and law enforcement are taken out of, what in most instances are, our self-serving and corrupt hands. This is our future. When has already been determined. Ancient peoples foretold this. One thing we can look forward to is knowing that some day after contact officially happens, a new governing instrumentality will be instituted on this planet by our overseers, the Atlanteans. This planet will be overhauled in a positive way, to our benefit, not to the benefit of self-serving interests. Presently, most of us live like idiots in this world at the hands of these self-serving interests.

Those who are self-serving and corrupt should put themselves on notice; they would do well to note that death is not the same for all, and in death there is no second chance. "Death" is an interesting "place," and some unscrupulous philosophies have offered bogus assurances of its accessibility to all. There are many misconceptions in this world. We need to broaden our picture; presently it is such a narrow one.

The overhaul of our planet will see the eventual introduction of the system of equalism, which is described in this book; the days of one individual having abundant wealth will be over. Individuals, corporations, and shareholders will only be able to earn a capped amount. Once this limit is reached, then the excess income will be pooled into a fund that will be at the discretion of the governing authority. You could say that the world's poverty problems will be fixed overnight when a true governing instrumentality is instituted on this planet and wealth is redistributed – that is, wealth is taken from the few and handed to the many, so that all are equal. Those days of the slaving worker struggling to make ends meet to fill the pockets of the shareholder, the obscenely paid corporate, and the like are coming to an abrupt end.

Those who are planning to do this already know what they are going to do and how they are going to do it. Strategies are in place. This is not fiction. This is our future, and not all that far away. There is certainly more to all of this. It will be interesting to learn of how the corresponding social and other changes that will occur are handled.

That a parent civilization far, far in advance of us should be considering contact now, in our generation, is the pinnacle of our closeted relationship, which has existed for over twelve thousand years.

THE END . . . AND NOW THE BEGINNING

Also by the Author

More information available at www.zodbooks.com

The First Cause
The Secrets of the Universe, the Brain, & Our Ancient Past
Finally Answered!
Volume I
ISBN: 978-0-9871167-0-3

An explosive book that solves the four greatest scientific challenges of mankind:
    The Origin of the Universe
    The Origin of Life
    The Workings of the Brain
    The Science of God

There is no other book like it on the market. This is the most important book ever written, one that is destined to change our present theories on everything from cosmology to spirituality. Described as a masterpiece.

Now On Sale at www.zodbooks.com

The First Cause

The Secrets of the Universe, the Brain, & Our Ancient Past Finally Answered!
Volume II

This book answers all the questions about Earth's ancient past.

Due for release in 2014

William Shakespeare – The Biography

A narrative non-fiction of the life of William Shakespeare.

This sensational true life story of William Shakespeare – once thought lost to us – not only exposes his tragic love affair that turned to hate and changed his world forever, but reveals the psychological torment that cursed him until his death. Discover a life story filled with comedy, romance, tragedy, and blackmail; one that had as many twists as the plots he wove in his plays.

Due for release in 2014

Piroshka

A narrative non-fiction of the life of Piroshka.

Due for release in 2015

Printed in Great Britain
by Amazon.co.uk, Ltd.,
Marston Gate.